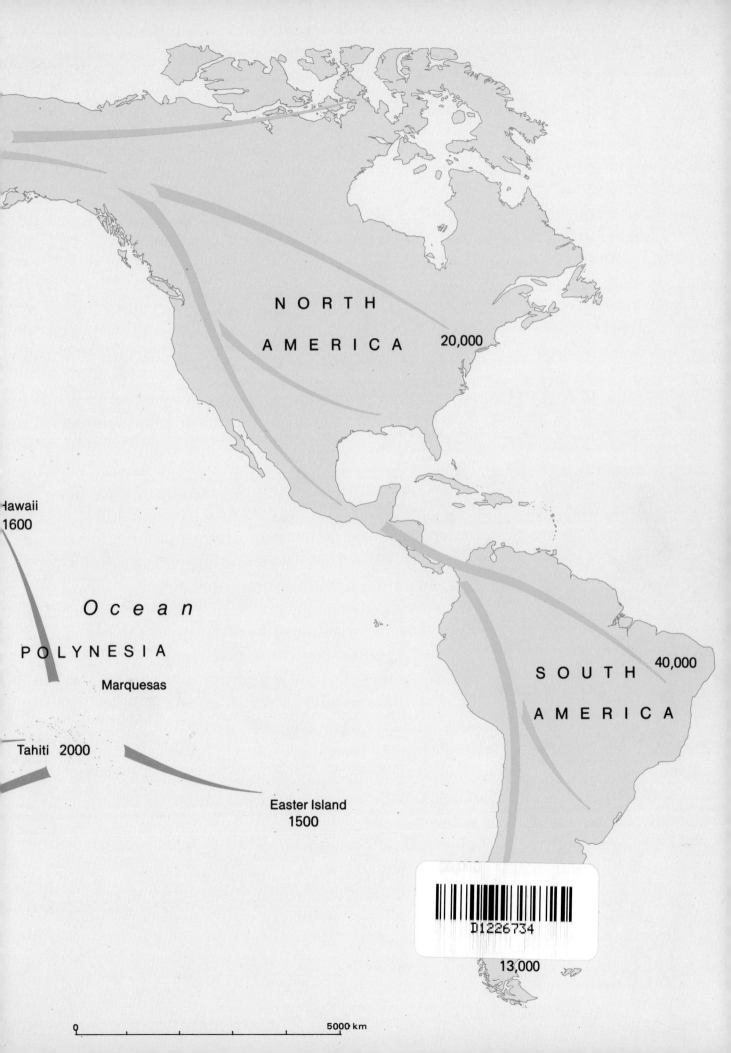

Hawaii
1600

Ocean

POLYNESIA

Marquesas

Tahiti   2000

Easter Island
1500

NORTH

AMERICA   20,000

SOUTH

AMERICA   40,000

13,000

0   5000 km

Man on the Rim

Around the Pacific Ocean, many
people still follow traditional
ways of life, especially those
whose subsistence comes from
the sea. *This fisherman belongs
to one of the communities of
'sea gypsies', who live over the
water off the coast of Sabah
(Borneo).*

Man on

the Rim

THE PEOPLING OF THE PACIFIC

Alan Thorne and Robert Raymond

ANGUS
& ROBERTSON
PUBLISHERS

ANGUS & ROBERTSON PUBLISHERS

Unit 4, Eden Park, 31 Waterloo Road,
North Ryde, NSW, Australia 2113;
94 Newton Road, Auckland 1,
New Zealand; and
16 Golden Square, London W1R 4BN,
United Kingdom

First published in Australia
by Angus & Robertson Publishers in 1989
in association with ABC Enterprises
(for the Australian Broadcasting Corporation)

A PIC venture

Copyright © Alan Thorne and Robert Raymond 1989
Designed by Harry Williamson

Cartography by Winifred Mumford
Logotype for TV series: Michael Berry

National Library of Australia
Cataloguing-in-publication data.

Thorne, A.G.
    Man on the Rim.

    Bibliography
    Includes index.
    ISBN 0 207 16246 8.

    1. Pacific Area — Civilization. 2. Ethnology — Pacific Area.
    1. Raymond, Robert, 1922-   . II. Title

909'09823

Typeset by Solo Typesetting, South Australia.
Printed in Australia by Griffin Press.

In Australia, hunter gatherers obtained their food with relative ease, and had a great deal of time for artistic pursuits and ceremony. *A dance group called Tjapukai performs in the rainforest near Cairns, in north Queensland.*

Contents

Introduction

In the last twenty years there has been a revolution in anthropology and archaeology. Not only has it become clear that the human species emerged on to the world scene much earlier than suspected, but we have obtained striking new insights into our social, cultural and technological development since then. On the evidence of fossil discoveries—many of them in unexpected places—we have a new perspective on such fundamental advances as the domestication of plants and animals, the discovery of metals, and the beginnings of ocean voyaging.

And the most surprising aspect of this story is that many of these findings have been made, not in Africa or the Middle Eastern 'cradle of civilisation', but in the long-neglected Eastern Hemisphere, in the lands surrounding the Pacific Ocean.

It now seems that the first true humans, *Homo erectus* (upright man), began to move out of Africa more than a million years ago, towards Europe and Asia. This expansion, the first great intercontinental migration of the human species, began the diversification of the major ethnic divisions of humanity. It also led to the occupation of the Pacific basin, which until then had been entirely uninhabited.

The first humans to see the Pacific were hunter gatherers, who arrived from the west and settled down in the jungles of Southeast Asia and the fertile valleys of China. Like their counterparts in Africa and Europe, they slowly developed, both physically and culturally. By about 100 000 years ago *Homo erectus* in Asia had become *Homo sapiens*—modern man.

Then began another great pulse of human migration, this time out of Asia. People moved south into New Guinea and Australia. They pushed north into Siberia, then across into North America and down into South America. And finally they spread right across the islands of the Pacific Ocean itself. In their long march from Asia to the tip of Tierra del Fuego those early explorers travelled halfway round the earth. It was the longest journey in the human occupation of the planet. Taken together, the migrations which settled the Pacific basin comprised the most adventurous

The symbol of the Pacific people is the outrigger sailing canoe, which over thousands of years of exploration and discovery enabled them to occupy the remotest corners of the world's largest ocean. *In island Southeast Asia people still use canoes like this for fishing and travelling.*

7

The great stone heads carved by the Olmec people of Mexico are among the most enigmatic works of art in all the Americas. *This is one of the collection of Olmec sculptures in Parc La Venta, in Villahermosa.*

phase in our history. And it was all over before the Western world even knew the Pacific existed.

Until quite recently, the true dimensions of these achievements were barely appreciated. The domination of Western culture and technology in human affairs over the past 500 years had one insidious side effect — it lulled the Western mind into an indifference to the Eastern Hemisphere, and to the civilisations of half the world's people. Even the rise of anthropology and archaeology, which began in Europe, tended to maintain a Eurocentric view of the world, based on the theories of human development that were defined there more than a century ago. As for the East's own awareness, that had been stifled by centuries of Western influence — cultural, political, and economic. Only now, as that huge reservoir of human energy and ideas begins to bubble again, is the evidence emerging of a truly remarkable cavalcade of human evolution.

Suddenly, we see the Pacific basin as a single human universe, from Tasmania right round to Tierra del Fuego, created by a common process of evolution and expansion that began in mainland Asia. Despite all the differences in faces, skin colour, art, music, language, foods and life-styles, there is a fundamental affinity, a sharing of origins, that binds them all.

It is this million-year chronicle of human development in the Pacific that we outline in our television series, 'Man on the Rim'.

We are fortunate in that in many parts of the Pacific basin cultural traditions have survived from the past with surprising tenacity. Thus we have been able to talk to hunter gatherers in Australia, Mayan weavers in Mexico, outrigger sailors in Melanesia, Arctic whalers in Bering Strait, 'sea gypsies' in Borneo, silk farmers in China.

In our exploration of this vast arena we have become very conscious of the role that ocean voyaging played in the settlement of the Pacific basin. Without maps or instruments, before boats as we know them were in use anywhere else in the world, those explorers crossed wide stretches of open sea, setting out to distant landfalls that were far beyond conceivable knowledge. In their boat-building, sailing and navigational skills the early Pacific peoples had no equal.

It has also become clear that the evolving cultures around the Pacific rim made, quite independently, two advances that were critical for human development. One was the transition from hunting and gathering to cultivating crops and keeping animals. In the process the Pacific people brought to the human menu an extraordinary array of foods: rice, corn, potatoes, tomatoes, beans, pumpkins, marrows, peppers, chillis, peanuts, soybeans, sugarcane, coconuts, avocados, bananas, pineapples, oranges and lemons, tea, cocoa, nutmeg, cloves and ginger, as well as chickens and turkeys.

The other crucial advance was the discovery of how to extract metals from their ores in the rocks, and the replacement of tools of stone, wood, bone and shell with those of bronze, iron and steel. In many critical areas of metallurgy it is now obvious that Eastern metalsmiths were far ahead of those in the West.

But alongside the belated recognition of such achievements, there is also a growing awareness that the early Pacific colonists had an environmental impact on those virgin lands, the magnitude of which is only now being realised. The regular burning of the bush in Australia and California, the clearance of the forests on Hawaii, Tahiti and Easter Island—these practices actually created the landscapes which the first European explorers found so 'natural'. The implication of early man in the extinction of many large animals during the ice age has long been suspected, but the widespread modification of environments brought about by hunter gatherers has been one of the great unrecognised consequences of the peopling of the Pacific.

Today, as world attention turns towards the Pacific basin, and research into its prehistory accelerates, intriguing questions multiply. If, as the evidence suggests, the Americas were settled from Siberia across a land

The first settlers of the northwest coast of North America were skilled wood-carvers, and dance masks were among their most striking works. *This mask is in the Museum of Anthropology of the University of British Columbia in Vancouver.*

In China, despite the decline of past centuries, and the upheavals of more recent times, a few classic buildings survive amid growing Westernisation. *The Temple of Heaven in Beijing dates from the Ming Dynasty, and is about 700 years old.*

bridge to Alaska, why are the earliest known sites of human occupation to be found in Brazil, and not in North America? If the Chinese had the compass, the sternpost rudder and huge ocean-going ships that reached East Africa hundreds of years before Columbus, why did they not explore the Atlantic? And why was Australia the only continent to miss out on the Agricultural Revolution and the Age of Metals?

For the traditional inhabitants of the Eastern half of the world, in the forty-five nations of the Pacific basin, recent finds in anthropology and archaeology are opening windows on a long-neglected past. In many places, as a result, cultures which were fading are not only surviving but gaining new strength.

For the immigrants into the Pacific of the last few hundred years—and the rest of the world—these new perspectives are a glimpse of the future. They define the matrix which will shape the next critical passage in human history: the Pacific Century.

1.
First Footsteps

The arrival of the first human beings on the western margin of the Pacific Ocean was one of the most momentous events in the history of mankind. It led to the conquest of the world's largest ocean, the occupation of four huge continents, and the diversification of the human species into some of its most numerous, energetic and colourful cultures.

And yet the story of that initial settlement of the Pacific basin is still little known or understood. The probable routes of migration have been inundated by the rising seas since the end of the ice age, 10 000 years ago. Some areas where the people first lived in the tropics of Southeast Asia are covered by dense rainforest, and others are now intensively farmed or occupied by teeming populations. All these factors make archaeological work difficult. The result is that the physical evidence of those first settlers is limited to a small collection of fossil bones and teeth.

To complicate matters, the habitable areas have varied dramatically in outline over the past million years, as the Pacific Ocean has risen and fallen from its present levels in step with the growth and shrinkage of the polar ice caps. Land masses and islands have been joined and separated, shorelines raised and lowered. Climates have changed, and with them the environments in which people had learned to live.

That the first human settlers in the Pacific basin were able to adapt their ways of life to all these variations, and expand into every corner of this hemisphere, is an enormous tribute to human resourcefulness. That capacity made the human species the most widely distributed mammal on the planet.

The Pacific story really begins in Africa, where humanity itself began, some 1.5 million years ago. Around that time there emerged, from a spectrum of closely related human-like creatures, or hominids, one particular species, *Homo erectus* (upright man), which was to become the progenitor of modern man, *Homo sapiens*.

Homo erectus was the first of those developing man-like creatures to

The modern people of Southeast Asia have roots which go back to Africa, to the emergence of the first true humans, *Homo erectus* (upright man). Bands of those early hunter gatherers arrived on the western margin of the Pacific Ocean more than a million years ago. *These children live in a village near the coast of Sabah (Borneo).*

exhibit what we know as the basic human form. The skeleton from the neck downwards would fit into the range of modern humans, and in this area there has been little change since then. Above the neck *Homo erectus* still had some way to go. The brain had a maximum capacity of just over 900 cc (which would put it close to the lower end of the range for living people), but the skull had a long, low crown, heavy brow ridges, a projecting face and large teeth.

Homo erectus moved ahead of other hominids by his use of increasingly sophisticated tools. Crude stone tools had been used as early as two million years ago—notably by *Homo habilis* (handy man)—and knowledge of how to make tools was spreading. But the hand axes found in campsites used by *Homo erectus* were quite well-made, and may have been used for more than simply chopping meat or smashing bones to obtain the marrow.

Homo erectus also displayed one other initiative which was to mark a significant milestone in human evolution, and put all who came thereafter on a plateau above other inhabitants of the planet. Some time between one and 1.5 million years ago these people began to leave the environments where they had evolved, and move—not just to nearby areas, but out of the African continent itself, north into Europe and east towards Asia.

By that expansion of the human geographic range into vast, previously unoccupied continents, the human species passed a critical evolutionary threshold. While other hominids in Africa had been limited by the constraints of their environments, and their capacity to deal with them, *Homo erectus* discovered how to adapt to new and unfamiliar environments, and to master them rapidly.

And that spirit of curiosity, persistence and unfailing adaptability, which overcame man's earlier, instinctive fears of the unknown, was to display itself in its most vigorous form once *Homo erectus* had finally reached the tropical lands on the margins of the Pacific Ocean. After nearly a million years of learning and adapting, the descendants of *Homo erectus* were to get on the move again, this time bursting out of Asia in a series of even more adventurous migrations to the farthest limits of the Pacific basin, occupying and settling an area covering more than half the planet.

Ironically, our first knowledge of *Homo erectus*, the pioneers of human exploration, came not from Africa but from Southeast Asia. And that discovery arose out of one of the most fascinating and romanticised preoccupations of the nineteenth century—the search for the 'missing link'.

Few intellectual concepts have ever aroused as much controversy and speculation as Charles Darwin's theory of evolution, as set out in 1859 in *The Origin of Species*. The conclusions reached by Darwin from observations made on his five-year voyage in the *Beagle* were to transform our way of thinking about the human past.

Darwin had three inspired insights. First, he saw living things not as static, immutable, one-off creations, as the Bible dictated, but as organisms in a constant state of change, with some species dying out while others came into existence. Second, he suggested that all groups of living things were descended from a common ancestor, but had diversified through change and adaptation. And finally, he proposed that the mechanism of change was natural selection—what came to be known as 'the survival of the fittest'.

These ideas reverberated throughout universities, scientific societies and other learned bodies of the Western world—especially when they were supported by another English naturalist, Alfred Russel Wallace, who after many years spent in Southeast Asia had independently come to much the same conclusions as Darwin.

One young man who was caught up by this revolutionary wave of thinking was Eugene Dubois, who had been born in Eisden in Holland in 1858, the year before *The Origin of Species* was published. Dubois therefore grew up in a period when Darwinism was threatening to overturn classical theories of geology and biology, and although his family was conventional and religious (his sister became a nun), he was encouraged by his father, a pharmacist, to take an interest in science. As a child he

The first true human, *Homo erectus*, had a smaller braincase and more prominent brows and face than modern people. Below the neck, however, the skeleton was very similar to ours, and little change has taken place in the past million years. *Model in the Java Man Museum at Sangiran, in Java.*

began to accompany his father on expeditions near his home, collecting animal fossils (as young Darwin had collected plants and butterflies). Then, while at high school, Dubois had an experience which shaped his life; he heard a talk by a distinguished German scientist on the new theory of mankind's evolution from prehistoric apes.

By then, many thinkers had begun to accept the general principles of evolution—that modern species of animals and plants were descended from earlier, more primitive ancestors—but the application of this theory to the human species was harder to swallow. The biggest problem was the lack of evidence. Where other animals were concerned there were at least some fossils to show a possible line of descent. For man there was virtually no fossil record, and certainly no connection with what appeared to be our nearest relatives, the great apes.

Despite the speculative basis of Darwin's theory, and the derision which it attracted from creationists ('Is it through your grandfather or your grandmother that you claim your descent from a monkey?', Bishop Wilberforce asked Thomas Huxley in their famous debate at Oxford in 1860), Dubois became an ardent evolutionist. But first he had to think of a career, and was persuaded by his family to enter medical school in Amsterdam. After graduation, however, he lost interest in the idea of practising medicine, and in 1886 obtained a university lectureship in anatomy.

Academic life gave Dubois time to think about the topic which had never left his imagination—human evolution. It also provided the opportunity to study the only evidence which had even the remotest bearing on this question: an incomplete skeleton found in a cave in the Neander Valley near Dusseldorf in Germany in 1856.

The bones of Neanderthal Man, as he came to be known, were obviously human, but sufficiently different in form and development from those of modern people to start a lively controversy. Some experts thought they must have belonged to an imbecile, or a diseased individual—or even to one of Napoleon's deserters. Others, however, recognised the antiquity and perhaps ancestral significance of the bones—and Dubois finally came down on this side of the argument.

To Dubois, Neanderthal Man suggested that there must exist even earlier and more primitive creatures, closer to the point where apes and humans had separated on their evolutionary paths. And so he began to think about a 'missing link', which would finally validate the evolutionary nature of human origins. But where to look for it? The most likely area seemed to be the tropics, the home of man's nearest relatives—gorillas, chimpanzees, gibbons, and orang-utans. Darwin had suggested that the earliest ancestors of man had lived in some 'warm, forest-clad land'.

Alfred Russel Wallace had also believed that human forebears might be found in the tropics. In his natural history, *The Malay Archipelago*, published in 1869, Wallace had described the only Asian great ape, the orang-utan, which was found in Sumatra and Borneo. He thought it 'very remarkable that an animal so large, so peculiar, and of such high type of form as the orang-utan should be confined to so limited a district'. Wallace

Eugene Dubois, the Dutch
anatomist, who discovered the
first fossil of *Homo erectus* at
Trinil in Java in 1891, had been
inspired by Darwin's *The Origin
of Species* to search for the
'missing link'. *Dubois could
raise no support for his work,
and got himself to the Dutch
East Indies by joining the army.
Dubois is seen here, second from
top, on the voyage east.*

went on to speculate: 'With what interest must every naturalist look
forward to the time when the caves and tertiary deposits of the tropics may
be thoroughly examined, and the past history and earliest appearance of
the great man-like apes be at length made known'.

Dubois was excited and inspired by Wallace's suggestion—particularly
as the area in question was, conveniently, part of Holland's possessions in
Southeast Asia. He decided to give up his teaching and search for the
'missing link' in the Dutch East Indies. His colleagues at the University of
Amsterdam tried to dissuade him from such a risky undertaking, especially
since he had recently married, and his wife was expecting their first child.
Such arguments seemed justified when Dubois failed in all his attempts to
obtain government support for his plan. But he persisted, resigning his
university post and enlisting as a military surgeon in the Dutch East Indian
army. In October 1887 Dubois, with his wife and baby daughter, sailed in
the *Princess Amalia* for Sumatra.

Dubois was assigned to a small hospital at Pajokumbu, in the interior,
and found himself with plenty of time to explore and search for fossils. He
discovered and excavated many limestone caves and deposits in the
Sumatran jungle. On one occasion he forced his head and shoulders into a
low cave and found bones everywhere—but they still had shreds of
decomposing meat on them. Dubois realised that he was in the den of that
formidable predator, the Sumatran tiger. In his haste to back out he
wedged himself into the opening, and his local guides had to drag him out
by his boots.

Despite all his searches, over a period of more than two years, Dubois
found no human fossils of any interest. In 1890, after a severe attack of
malaria, he was transferred to Java, and placed on the reserve list.
Although by now less hopeful, he persuaded the colonial administration to
supply him with convict bearers and labourers, and set out to search again.

In eastern Java Dubois found a much more promising environment. Here
the landscape had been lifted up in many places along the still active
volcanic belt which passed through this part of Indonesia. Erosion from the
heavy monsoon rains and the swift-flowing rivers had cut down through
the sediments, exposing rich beds of animal fossils, some dating back a
million years or more. The area was dotted with villages, and as the
farmers worked their paddy fields after the rains they constantly came
across the bones of extinct animals: rhinoceros, a primitive elephant
(*Stegodon*), giant buffalo, pigs, tigers, tortoises and deer. Dubois set teams
to work in a number of places.

Dubois himself set up camp beside one of the most promising sites: the
high, silty bank of the Solo River, near the village of Trinil. In August 1891,
after the river had gone down at the end of the monsoon, he began to
excavate a trench right down through the thirteen-metre-high bank. In
September he made his first really significant find. It was only a single
tooth—not quite human, nor exactly ape-like—but it was obviously very
old. Then, in October, one of Dubois's workmen found, only a metre away,
a heavy object that looked like a small turtle shell. Carefully scraping away

the encrustation, Dubois realised that what he was holding was a skull cap—the top of a cranium. But despite his anatomical knowledge he could not decide what kind of animal it came from.

Dubois continued with his excavations at Trinil, and slowly accumulated a few more fragments of bone: a piece of a jaw, a human-like thigh bone, another tooth. Dubois studied his finds for nearly a year, building up in his mind a picture of the creature that they came from. At last he realised that it was just what he had come to Southeast Asia to find: something with a fairly small brain, intermediate between apes and humans, but with a leg bone that indicated that it walked upright, like a human. Dubois cabled his friends in Holland that he had found 'the missing link of Darwin'.

When Dubois published his discovery in 1894 he called it *Pithecanthropus erectus*—upright ape man.[1] What came to be known as 'Java Man' was immediately recognised as being of enormous significance. Here, in far-off Java, was the first specimen of man's immediate ancestors—the first true human. Here was the first solid evidence that humans had an evolutionary past, as Darwin and Wallace had predicted.

Since that first find there have been other discoveries of *Homo erectus* in Java, especially in the area around Sangiran, a few kilometres north of the city of Solo. In the 1930s and 1940s a German palaeontologist, G. H. R. von Koenigswald, found several skull caps and jaw fragments. Since Indonesia became independent in 1949 the research has been continued by locally trained specialists, in particular the anatomist Professor Teuku Jacob, of the University of Gajamada in Jogyakarta. The collection now includes fragments of faces, braincases, teeth and limb bones from forty individuals—a third of all *Homo erectus* fossils found worldwide. It includes the remarkable 'Sangiran 17', discovered in 1969. This has the face attached to the braincase, and was the first virtually complete *Homo erectus* skull ever found.

The nature of the sediments around Sangiran in which the human fossils were found, and the range of other animal and plant fossils they contain, indicate that at the time these early humans settled here this locality, which is now well inland, was on the north coast of Java. The vegetation included mangroves, pandanus and other palms, and the layers of marine mud contain oyster and other shells, as well as sharks' teeth. Some of the human fossils were found in the marine and estuarine sediments, so it appears likely that these early hunter gatherers were living on the coast itself, or very close to it—perhaps on a river leading to the ocean. This makes sense, because the junction of two habitats—in this case the coastal forest and the sea—is always a good place to find animals. The combination of marine and land environments is particularly productive.

This picture of life around Sangiran also hints at the possibility that the first arrivals in Southeast Asia might have followed a generally coastal route from Africa, along the shores of Arabia, India and what is now the Malaysian peninsula. The coastal corridor would have provided a continuous and familiar environment for travellers to follow, minimising the need to conquer and adapt to new and strange territory. Of course, while

The Solo River, in Java, at the place where Dubois made his historic discovery of *Homo erectus* (later known as 'Java Man'). *Alan Thorne holds a cast of the cranium; the original is kept in The Netherlands.*

some groups of *Homo erectus* were continually moving on, heading east towards Southeast Asia, others may have stayed and expanded inland, into what is now Iran, Pakistan, and India. And some, as we shall see, certainly crossed or rounded the Southeast Asian peninsula and headed north.

Once settled in Java, *Homo erectus* maintained a settled way of life, with remarkably little variation, for more than half a million years, despite slow but significant changes in the environment. Around Sangiran, the geological and fossil record shows that the land was gradually uplifted. The estuarine vegetation turned into forest, interspersed with patches of grassland. Evidence of volcanic activity suggests that the landscape was not unlike much of Indonesia today, with many lava-strewn and ash-drenched areas, as well as rainforest in places.

The animals of that time included many species that are familiar today, except that some of them were much larger. There were several species of a primitive elephant, and a giant water buffalo, with a horn spread of three

Around the area of Sangiran, in Java, rivers fed by the monsoon rains are eroding ancient marine sediments, exposing bones of *Homo erectus* and extinct animals that date back more than half a million years. *At that time, Sangiran was on the north coast of Java. Today, because of volcanic uplift, it is far inland, near the city of Solo.*

Since Eugene Dubois's original discovery in 1891, bones from a total of forty individual *Homo erectus* people have been found in Java—a third of all the 'upright man' fossils found worldwide. The prominent brows are characteristic of these early humans. *The collection is held at the University of Gajamada in Jogyakarta.*

metres. A huge land tortoise, with a shell two metres long, died out soon
after human fossils appear in the sequence; it is intriguing to wonder
whether *Homo erectus* had anything to do with this.

There is only one living survivor of those times—*Varanus komodoensis*,
the giant lizard known as the Komodo dragon. This black, heavily built
member of the worldwide family of monitors (which includes the goannas
of Australia) grows to a length of more than three metres, and thus greatly
exceeds all other lizards. It is a powerful carnivore, catching deer, pigs,
and other mammals, and has been known to take unwary humans. It was
once widely distributed in the Indonesian islands but is now found only on
three volcanic islands, including Komodo, east of Bali. It would have
proved a formidable quarry for *Homo erectus* hunters—especially as we
have no certain knowledge of what kind of weapons and tools they used.

In fact, apart from the certainty that the first Javan people were hunter
gatherers, we know virtually nothing about how they lived. The sites

Komodo, near Bali, shows the typical eroded volcanic landscape of so many of the islands of Indonesia, which lie along the Pacific's 'ring of fire'. *The thin covering of green vegetation on the higher slopes is only seen after the monsoon; generally the island is brown and barren.*

Komodo dragons, which grow to a length of more than three metres, are the world's largest lizards. These predators catch deer and pigs, and have been known to take an unwary human. *Once widespread in Indonesia, they are now found on only three islands, including Komodo.*

Homo erectus people, who were hunter gatherers, lived in Java for more than half a million years with little change in their physical characteristics. Their diet was mainly vegetarian, with meat when they could kill an animal with their rudimentary weapons. *Models in the Java Man Museum at Sangiran.*

where their remains have been found are not living areas, where people camped or buried their dead, so we have no evidence of the size of their groups, or of the proportions of men, women and children. And although stone tools have been found in many places in Indonesia, including Java, none of them has ever been associated directly with *Homo erectus*. Even those found in proximity to the human bones were probably washed down from more recent levels, and became mixed with the older fossils. Despite long and careful searching, archaeologists have never found any signs of a stone industry, such as the flakes and cores associated with tool-making, which could be linked to the early Indonesians.

Given that a third of all *Homo erectus* fossils known worldwide come from Java, and that everywhere else — in Africa, Europe and East Asia — *Homo erectus* remains are associated with an abundance of stone tools, this absence of tools is quite odd. It may be that Javanese people of this period used other materials to make their utensils and tools. One of those materials, to judge from life in Java today, might well have been bamboo — at least for cutting edges — since the tough outer skin can be shaped into a razor-sharp blade.

Anatomically, there seems to have been remarkable stability in the Javan people over that half-million years. In build they were short but quite muscular, as you might expect in mobile hunters. The general skull shape persisted, with a receding brow leading back to a low crown. The skull itself was thicker than in modern people (and even thicker than in other specimens of *Homo erectus*, found elsewhere). The face was large and very broad, under heavy brow ridges. The nose was flattish, and the lower part of the face projected.

The long occupation of Java by *Homo erectus* appears to have ended between 150 000 and 200 000 years ago. A set of human remains of about this age was found in the 1930s at a place called Ngandong, on the Solo River not far from Trinil and Sangiran. There were twelve partial brain-cases — unfortunately all missing the face — and some fragments of lower limbs, from men, women and children. Although obviously related to the other early Javanese, these people had larger brains, and showed more modern features of the brows and back of the head.

The Ngandong people show clear indications of development towards modern man, *Homo sapiens*. But there the line stops. There is no evidence of any human populations in Java (or anywhere else in Indonesia) after the Ngandong period, until the end of the ice age, some 10000 years ago. The people who begin to appear then in the fossil record are not at all like Java Man. They were the product of a quite different line of development of *Homo erectus*, that had taken place far to the north of Java, in Asia itself.

The discovery of Java Man, the first specimen of *Homo erectus*, was strong evidence in favour of Darwin's theory of evolution, but obviously much more was needed to convince the sceptics. In fact Dubois himself, after his return to Holland in 1895, became so annoyed at criticism of his material that he withdrew it from all examination. Only in 1923, after an appeal to the president of the Dutch Academy of Sciences, was Dubois persuaded to invite Ales Hrdlicka, Curator of Physical Anthropology at the US National Museum in Washington, to see the original bones.

So, in the opening decades of the twentieth century, the search for more examples of the 'missing link' gathered momentum in many parts of the world.

Africa was an obvious place to look, if only because of the existence there of the great apes, gorillas and chimpanzees, which were clearly, if distantly, related to man. But the next major discovery did not come until 1924, when Raymond Dart, Professor of Anatomy at the University of Witwatersrand in Johannesburg, announced that a skull found in a limestone quarry was that of an upright-walking creature that he named *Australopithecus*. It was closer to humans than to apes, but older and more primitive than Java Man—nearer to the 'missing link'. Dart's opinion, too, was challenged, but finds in many parts of Africa over the past sixty years have vindicated his judgement, and firmly established that continent as the home of the family of man.

Meanwhile, because the first specimen of *Homo erectus* had been found in the eastern jungles, the search for early human fossils spread to other parts of Asia, and particularly to China.

In China fossils were to be found in large numbers—not by digging, but by simply going to the traditional pharmacies and asking for 'dragon bones'. This was the name given to the many different kinds of fossils, dug out of caves and limestone deposits all over the country, that were ground up to make prescriptions. The 'dragon bones' on display often included fragments of extinct tigers, pandas, crocodiles and deer, side by side with the dried skins and organs of a bizarre selection of modern mammals, reptiles and amphibians. It was this unlikely scene which provided the first clue to one of the most famous archaeological discoveries of all.

In 1899 a European doctor noticed an unusual tooth in a pile of bones about to be ground up in a pharmacy in Peking. The tooth found its way back to Sweden, where it was first dismissed but then, some years later, identified as that of either an unknown ape or a very primitive human. But it was not until 1921 that a Swedish geologist, John Gunnar Andersson, went to China and tracked down the location where the tooth had been found. This was a steep, rocky ridge known as Dragon Bone Hill, near the village of Zhoukoudian, forty-eight kilometres south of Peking. The limestone hill was riddled with fissures and caves, many of them choked with sediments containing a rich assortment of fossil bones of animals which had apparently died or fallen into sinkholes.

Andersson began digging out fossils and sending them back to Sweden for study. He was soon joined by another foreigner in China, a Canadian named Davidson Black, who was head of the Department of Anatomy at the Peking Union Medical College. Black had become interested in the question of man's origins while studying anatomy in England during World War 1;

The basin of the Yellow River in China was the scene of some of man's earliest organised activities, as people began to live and hunt together in groups. Over the past 8000 years it has supported large populations and highly developed civilisations. *The name Huanghe, 'yellow river', comes from its immense load of fine silt, which it deposits to make fertile plains.*

his tutor had been working on the famous 'Piltdown Man' skull (later proved to be a fake). Black was convinced that some form of primitive man would one day be found in China, and was determined to be the one to recognise it. He had already searched without success in north China and Thailand, but was more hopeful about Dragon Bone Hill, even though the finds of anything resembling human remains were rather meagre.

Among the fossils eventually sent to Sweden by Andersson was another tooth, similar to the one found earlier. It was at first thought to be that of an ape, but the European experts finally conceded that perhaps it was human. When this report reached Peking in 1926, Davidson Black felt confident enough to announce to the scientific world that a new species of primitive man had been discovered in China. He named it *Sinanthropus pekinensis*, 'the Chinese man of Peking'.

Black's two papers in the British journal *Nature*[2] set in train a series of events which was to make Dragon Bone Hill, the home of Peking Man, one of the most famous archaeological sites in the world.

At first Davidson Black's claim, based as it was on a couple of teeth, was received with some scepticism. But Black persuaded the Rockefeller Foundation to finance large-scale excavations at Zhoukoudian, and soon an army of workers began digging a huge pit into the top of the hill. The work revealed that a chain of interconnected caves had once extended through the limestone core of the hill, from the summit down to the lower level, opening into an immense cavern 140 metres long and 40 metres high. Over perhaps a quarter of a million years the main cavern and upper caves had gradually filled with rubble and sediments, as floods and rains washed material into them.

What the excavators also discovered was that over that same immense period of time the main cave showed signs of an intermittent but persistent sequence of human occupation. But the fossils were locked into a matrix of rocks and pebbles, cemented together by dissolved and redeposited lime-stone, and the workers had to use blasting powder and steel drills to make any progress at all. There is an unusually graphic record of this excavation, in photographs taken by a young Chinese archaeologist, Jia Lanpo, some of which we have reproduced.[3]

The first discovery to support Davidson Black's claim came soon after the excavation began, in 1928. It was only a small fragment of a jaw, but it encouraged Black and the Chinese archaeologists directing the monster dig. Next year came indisputable confirmation of their belief in the significance of the Zhoukoudian site—a skull cap of *Homo erectus*. It was found by Pei Wenchun, later to be a senior researcher at the Institute of Vertebrate Palaeontology and Palaeoanthropology in Peking. A worker on the site has described the events:

'It all took place after 4 o'clock in the afternoon. We had got down about 30 metres deep and only three men could stand at the bottom of the hole. It was there the skullcap was sighted, half of it embedded in loose earth, the other half in hard clay. The sun was almost set and the light getting poorer. The team debated whether to take it out right away or to wait until the next day when they could see better. The agonizing suspense of a whole night was felt to be too much to bear, so they decided to go on . . .'

Human fossils continued to come out of the excavation, which eventually hollowed out the whole interior of Dragon Bone Hill. By 1937 skeletal parts of more than forty men, women and children had been unearthed, all from this single cave site. The recovered fossils also included a multitude of animal bones — many of them clearly the remnants of meals. These were found with the human bones, scattered through the many different layers of sediments that mark the enormous period of time that Peking people lived here.

a

b

c

d

(a): The great excavation at Dragon Bone Hill, at Zhoukoudian near Beijing, was begun in 1927 under the direction of a Canadian doctor, Davidson Black. *(b):* The huge pit was sunk straight down in the hill, with each section to be dug carefully marked out and numbered. *(c):* The archaeological team eventually reached the bottom of the large cavern, the Lower Cave, which contained the fossils of 'Peking Man'. *(d):* A Chinese archaeologist, Jia Lanpo, extracting fossils from the rock-hard debris which had filled the Lower Cave. *Jia Lanpo, who is the sole survivor of the original team of excavators, took these photographs between 1927 and 1937.*

Recent research at Zhoukoudian by a team of Chinese scientists, including geologists, palaeontologists, and specialists in the latest dating techniques, has confirmed that people lived in the so-called Lower Cave over a period of something like a quarter of a million years, from around 500 000 years ago until about 250 000 years ago.

This occupation was not continuous, because the great cavern was not always open to the outside. Sometimes both lower and upper entrances were closed by rock falls or accumulations of sediments. The geological evidence shows that the cave was open, and occupied, for thirteen main periods over that quarter-million-year span. Nevertheless, it has provided us with the longest sequence of early human occupation available anywhere, and the skeletal remains constitute our most comprehensive record of a critical period in human evolution.

Although Peking Man was clearly *Homo erectus*, like Java Man, there was one striking difference in the context of their existence: at Zhoukoudian there were stone tools everywhere—more than 100 000 of them. They show a variety of shapes and sizes, similar in general to the stone tools used by our much more recent ancestors, and even by some modern hunter gatherers. The design of some of the stone implements suggests that they were used to make other tools of wood or animal products, such as bone and shell. Overall, the tool assembly of Peking Man is an indication of increasing intelligence and industry.

An even more convincing sign of evolutionary development was the discovery that from quite early in the occupation of the Lower Cave the people there kept fires burning for long periods. In places there are thick beds of dark ash—one more than sixty centimetres deep. The fires were all restricted to what were unmistakably hearths, and some contain lumps of baked, heat-cracked clay, like fire-bricks. There is nothing to indicate what the fires were used for, no firesticks to show how fire was made, and no evidence that these people in fact knew how to make it. But there is no better evidence anywhere for man's first deliberate use of an enormously potential tool.

The Zhoukoudian people were hunter gatherers, and during the tens of thousands of years that they lived in this area they had to adapt to quite marked variations in climate, and the effects on their food resources. The overall change was from a warm, almost tropical climate to the present alternation of very hot summers and very cold winters.

The country around Zhoukoudian contained a good variety of well-stocked animal habitats, which may be one reason for the extremely long human occupation of the cave. There were densely forested mountains, grassy plains, and rivers and freshwater lakes. In the forests lived sabre-toothed tigers, leopards, bears, dogs, wolves, monkeys and rabbits. On the plains roamed large mobile animals such as elephants, buffaloes, rhinoceros, deer, ostriches and hyenas. Seasonality in hunting and foraging for berries, fruits and herbs is reflected in the changing pattern of bones and food remains over the year.

Anatomically, the Zhoukoudian people were very similar to *Homo erectus*

One of the lumps of baked clay found with fire-blackened hearths in the Lower Cave at Dragon Bone Hill. This evidence of the human use of fire is the earliest known anywhere in the world. *This and other fossils found in the cave are on view at the museum at the Zhoukoudian excavation site, forty-eight kilometres from Beijing.*

people in Java. They had the same general features of head shape, brow ridges, and facial proportions. This is a good indication of the wide original spread of *Homo erectus* in Asia. But a more detailed comparison shows a number of significant and interesting differences. The upper incisor (front) teeth of the Peking people are hollowed out at the back, or shovel-shaped—something not seen at all in Java Man. In addition, the face and brows are more delicate, the forehead more rounded. Taken together, and traced through the quarter-million-year sequence of Peking people, all these subtle variations in skull shape add up to an increasing modernisation of form, a trend towards a form more like *Homo sapiens*.

There are many reasons why the cave of Peking Man has excited the imagination of people everywhere. But there is one overriding consequence of the discoveries made there, which has a significance far beyond China and Java. With the sheer mass of evidence from Zhoukoudian we can be sure that the inhabitants of the cave were unquestionably human, and that those humans had slowly changed, physically and culturally, over something like a quarter of a million years.

In other words, they had evolved—and evolved in the direction of modern people, *Homo sapiens*. This was a massive body of support for the theory of evolution. After the discoveries made at Dragon Bone Hill no one could seriously doubt that humans, like other animals and plants, had reached their present form through the same process of change and adaptation.

Davidson Black died in Peking in 1933, but excavations at Zhoukoudian continued. The German anatomist Franz Weidenreich took over from Black at the Peking Union Medical College, and under his supervision the excavations added considerably to the Peking Man fossil collection. Then, in 1937, tension developed between China and Japan over the occupation by the Japanese of the province of Manchukuo, and Japanese forces established themselves on the very outskirts of Peking. All work at the Zhoukoudian site halted.

Weidenreich spent the next three years studying the fossils and making casts of them for overseas institutions. His meticulousness turned out to be an extraordinary bonus for the study of human history, for the entire Peking Man collection was about to disappear in a most baffling way.

When war broke out in Europe, and seemed likely to spread to the Pacific, the Peking Union Medical College feared for the safety of the Peking Man fossils. Arrangements were made to ship them to the United States, guarded by US marines from the embassy in Peking. Early in December 1941 they were carefully packed and put on a train to the port of Chinwangtao, where the American steamship *President Harrison* waited for them. But Pearl Harbor was attacked on 7 December, while the fossils, with their marine escort, were still on their way to the coast. In the confusion that followed the US marines were interned, the *President Harrison* was sunk, and somewhere along the way the crates of Peking Man fossils disappeared. They have never been seen since.

After World War 2 palaeoanthropological research in China was increas-

Part of the Lower Cave at Dragon Bone Hill, where 'Peking Man' people lived at different times for a quarter of a million years. *The opening looks out into the excavation that was dug down into the centre of the hill.*

ingly directed and carried out by Chinese archaeologists. Over the past forty years they have broadened their fields of excavation considerably, into all provinces of that huge country. The results have expanded their collection of human fossils into what is now the best and most continuous record for any part of the world, spanning as it does more than a million years. In 1987, at the Institute of Vertebrate Palaeontology and Palaeo-anthropology in Beijing, the staff set out on a long table the key elements of their entire collection, for us to film. To inspect this remarkable line-up was to see the evidence for a huge slice of human evolution.

The oldest human fossils found in China are two teeth from Yuanmou in Yunnan Province, in the extreme south. They are about one million years old. The teeth are upper central incisors, and they are distinctly shovel-shaped. This, as we shall see, is an important marker in the Chinese sequence.

Next in the chronology there was a skull and jaw from Lantian, in central

Part of the remarkable line-up of human fossils from China, extending back more than a million years, which was set out for us to film in 1987 at the Institute of Vertebrate Palaeontology and Palaeoanthropology in Beijing. *Alan Thorne is touching a model of what Peking Man is thought to have looked like.*

China, between 600 000 and a million years old. This skull is clearly *Homo erectus*, and is in fact very like the skulls found in Java. The resemblance supports the proposition that the initial settlements of eastern Asia and Southeast Asia were made by the one kind of people. Yet the Lantian skull is just different enough to show that groups in Java and China were beginning to diverge.[4]

Next, there was Peking Man himself, represented by a selection from the full range of fossils held in Beijing. Most of the skulls and limb bones were of course casts taken from the lost originals, but there were some actual specimens that had been excavated more recently. They show the slight changes towards modern people that were developing during the Peking Man period.

The changes became quite obvious in a nearly complete skull from Dali, near the ancient capital of Xian (where the Entombed Warriors were unearthed), and another from Yingkou in Manchuria. These two individuals are younger than Peking Man—between 150 000 and 200 000 years old. Their braincases are filled out, and although their brow ridges are still well developed the faces below these are smaller and more vertical, very similar to modern people. They show very clearly the transition from *Homo erectus* to *Homo sapiens*, and in fact most specialists put them in the latter classification.

Finally there were three finds which are unmistakably in the realm of modern humans. One was a delicate cranium from Ziyang in Sichuan that is about 30 000 years old. Another was a partial skeleton from Liukiang in the south, which is approximately the same age. And finally there were three well-preserved skulls from the so-called Upper Cave on top of the Zhoukoudian hill, above the Peking Man cave. These remains are all between 15 000 and 20 000 years old. They show that by 20 000 years ago there were modern humans living all over China (although some of the specific features of the present Chinese people had yet to appear).

At the end of the line-up in the Institute in Beijing there was the skull of a modern Chinese. Significantly, the upper front incisors were shovel-shaped—as are all the incisors in all the Chinese human remains going back a million years.

This feature, and in fact the whole assembly of fossils, is relevant to a major argument among anthropologists as to where modern humans originated. Some insist that it must have been in Africa, where the earliest humans appeared, and that from Africa they spread out across the world, as *Homo erectus* did much earlier. The Chinese evidence supports another argument, that the major human groups—African, European, Asian—developed more or less where they live today. Certainly, the Chinese sequence of fossils suggests that modern Asians evolved in Asia.

And it was from Asia that, some 50 000 or more years ago, those modern people began to move out into the whole Pacific basin. Their first great push was to the south—into the islands of the Philippines and Indonesia, including Java. Some went on, to discover and occupy a whole new continent: Greater Australia.

Excavations are still continuing at Dragon Bone Hill at Zhoukoudian, more than sixty years after the original discovery of Peking Man. *This work is taking place near the bottom of the hill, where excavators are recovering* Homo erectus *tools and animal fossils.*

2.
Casting Off

To understand and appreciate the astonishing series of human migrations out of Asia which began some time after 100 000 years ago, and which eventually reached the farthest shores of the Pacific Ocean, we must look for answers to a number of crucial questions: What was the southwest Pacific like at that time? What kind of people were those first explorers? What persuaded them to leave their homelands and cross oceans to unknown islands? And, finally, how did they do it?

The first and most critical call on the imagination is to picture the physical world of those early people, and to realise that it was nothing like the one we know. The sea levels that we see around our coastlines, the outlines of countries on the map that we find so familiar, the climates, landscapes and even vegetation that we associate with particular geographical areas, were all very different then. The reason is that over the past million years the earth has been through a number of glaciations or so-called 'ice ages', the last and most severe of which ended only 10 000 years ago.

It is still not clear what causes these global climatic cycles, in which very cold phases alternate with much warmer periods (such as the earth has been experiencing for the past few thousand years). Many factors may be involved: fluctuations in the output of energy from the sun; wobbles in the earth's rotation on its axis; even small variations in its orbit around the sun. Some cold phases may have been triggered by an increase in the reflectivity of the atmosphere, produced by vast dust clouds hurled into the upper atmosphere by volcanic eruptions.

Whatever the cause of the ice ages, they were marked by an overall lowering of world temperatures. This increased the amount of precipitation in the form of snow in the higher latitudes, causing the polar ice caps to expand. At the coldest periods of the ice ages the ice caps grew to several times their present size. In the Northern Hemisphere ice sheets many kilometres thick spread south across North America, Europe and Asia. In

The bamboo sailing raft may have been the craft which carried the first human migrants out of Asia into the islands of Southeast Asia, more than 50 000 years ago. *Alan Thorne tests an experimental raft off the coast of Java. The sail is made from two woven reed sleeping mats.*

Australia glaciers were common in the highlands of Tasmania, Victoria and New South Wales.

As all this water became locked up at the poles instead of finding its way back into the oceans, the sea level fell dramatically. It is surprising to realise that the present level, which we regard as normal, is in fact exceptional. The sea has been at this level for only the past few thousand years. Before that, for most of the previous 100 000 years, it was at least fifty metres lower, and during the ice ages even lower.

The most recent ice age was the coldest, and at its most severe phase, about 18 000 years ago, the sea was 150 metres, or more than 450 feet, below its present level. As the ice age came to an end and the ice sheets retreated the oceans rose again, reaching their present level only about 6000 years ago. (If the ice caps are melted still further by the 'greenhouse effect' the sea level will of course rise even more, inundating the coastal cities which have sprung up in the last few thousand years.)

A lowering of sea level of such magnitude had a dramatic effect on the map of the world, and nowhere more graphically than in the western Pacific. The South China Sea was reduced to less than half its present area. Many of what are now the islands of Southeast Asia—Sumatra, Java and Kalimantan (Borneo) in Indonesia, Palawan in the Philippines, and hundreds of other smaller islands—were part of the Asian mainland. To the north, Taiwan was a mountain range on a great plain extending from Vietnam to Japan, backed by the highlands of China. To the southeast, the floor of the Arafura Sea was exposed, linking Australia and New Guinea. Below Australia, Bass Strait did not exist, and Tasmania was a mountainous projection of the mainland. The combination of New Guinea, Australia and Tasmania is known as Greater Australia.

How is it possible to know when ice ages occurred, and what fluctuations there were in the sea level? The evidence for the duration and dating of the ice ages themselves comes from a variety of scientific investigations: the examination of cores of sediments drilled from the sea floor; the analysis of bubbles of air from past atmospheres trapped in existing ice sheets; the study of pollen from soil deposits on land. The geological record also provides convincing evidence of past events, and changes in sea levels in particular are often reflected in the coastal cliffs, wave-cut platforms and frontal dunes of present shorelines.

One of the most spectacular records of dynamic sea level changes can be seen on the Huon Peninsula, in northeast New Guinea. Here the land rises steeply from the beach in what appears to be a series of terraces or giant steps. These are in fact fossil coral reefs, formed at various times in the past when the sea level was very different from what it is now.

Reef-building corals are very active in the tropical waters around the Huon Peninsula, and a fringing reef is today being built just off the beach. Because the reef-building coral animals need sunlight to produce their limestone skeletons, reefs grow only in the surface layers of the ocean. Thus they mark the existing sea level. If the sea level rises or falls, and stabilises for a period, a new reef begins to grow at the higher or lower

N

level. So each fossil reef on the side of the Huon Peninsula represents a different sea level in the past.

This kind of record undoubtedly exists at many places in the Pacific, but elsewhere the rise of the sea to its present level has submerged the old reefs. They are visible on the Huon Peninsula today only because this part of New Guinea is very active geologically. For hundreds of thousands of years it has been slowly rising, at a rate of about four metres every thousand years. This uplift has exposed a whole sequence of ancient reefs high above the present sea level; the topmost reef is more than 200 metres above the beach.

Australian geologists have studied these reefs in great detail, and by taking a sample of coral from each they have been able to date them, using the carbon-14 count for recent reefs and the potassium-argon relationship for older ones. By correlating the dates of the reefs with the known rate of uplift of the Huon Peninsula it is possible to produce a graph of sea levels

Fossil coral reefs on the Huon Peninsula in New Guinea provide striking evidence of changing sea levels during past ice ages. As the polar ice caps grew or melted the sea level fell or rose; each time the level stabilised a new reef developed. *These reefs, some dating back 300 000 years, are exposed today because the Huon Peninsula rises about four metres every 1000 years.*

going back more than 300 000 years. It is a remarkable experience to climb this giant staircase to the base of the highest reef and to look down over nearly half a million years of geological history to the present coastline and its growing reef, far below.[1]

We will never know the full impact of these slow but far-reaching environmental upheavals on the evolution of the early people on the shores of the Pacific Ocean, and how they affected the occupation of different areas. The lowering of sea levels and the exposure of vast areas of sea bed were certainly the most obvious effects of the ice ages, but they were only part of a series of global environmental and climatic changes. Many of them are still poorly understood, or can only be guessed at—such as the variations in tides, sea currents and winds that must have occurred. There were probably significant alterations in the timing and intensity of the annual monsoon season, as well, and this would have produced significant changes in weather patterns from China down to Australia.

The Huon Peninsula also indicates, by the shell middens eroding out of the dunes above the beach, that it was on the migration route of people moving out of Asia towards Australia. *Shell deposits like these are the product of thousands of years of feasting on mussels, clams and other shellfish.*

Not surprisingly, the best picture we have of the environmental changes that ice ages bring about comes from the most recent glaciation, which began some 45000 years ago, and reached its most intense stage about 18000 years ago.

For lowland Southeast Asia (which, as already mentioned, then covered a much larger area) this was a time of lower temperatures—on average, two to three degrees Celsius lower—and lower rainfall. There was less rainforest, and the exposed sea floor was probably covered with grassland.

In New Guinea, with its central ridge of high mountains, the effects were dramatic. In the mountains the average temperatures were as much as seven degrees lower. The snow line was more than 1000 metres lower, and there were glaciers in the high valleys which lasted until about 10000 years ago. Below the cold uplands there was a vast belt of grassland, which survives only in scattered patches in what is now West Irian. Surrounding that there was a much narrower band of the kind of tropical rainforest so typical of New Guinea today. Finally there was a coastal fringe of the kind of dry, open woodland that occurs now much further south, in Arnhem Land in northern Australia.

Despite the fusion of many of the islands in Southeast Asia to the Asian mainland, there were always some water gaps between Asia and Greater Australia. One of these ocean trenches separates Bali in the west from Lombok in the east, in the chain of Indonesian islands, and extends north between Kalimantan (Borneo) in the west and Sulawesi (Celebes) in the east. This deep channel—which even at the time of the lowest sea level was never less than about seventy kilometres wide—helped to create one of the great biological boundaries in this part of the world. It marks a separation between the Asian zoogeographical zone in the west—with its typical animals, including elephants, rhinoceros, tigers, deer, monkeys and orang-utans—and the Greater Australian zone in the east, with its typical marsupials, monotremes (platypus and echidna), and birds of paradise.

This boundary was first noticed by Alfred Russel Wallace, the English naturalist, and was eventually named Wallace's Line. Instead of a line it is now recognised rather as a belt where the two zones overlap (Wallacea). In Sulawesi, for example, there is both an Asian monkey and a cuscus (marsupial possum). But in general, over millions of years, there was virtually no crossing of that boundary, even when the lower sea level greatly expanded the Asian zoogeographic zone. The few exceptions were birds, bats and insects which flew across the water gaps, some rodents and lizards which drifted across on logs, and plants whose seeds floated across.

It was left to a new kind of animal to transcend that natural biological boundary—people coming out of the north, with some form of watercraft. Who were these people, and what drove them on, during this time of great change and instability?

One of the myths of human evolution, perpetuated now for more than a hundred years, is that since the emergence of *Homo sapiens* there has been a steady increase in the level of human intelligence. The implication is that people of the past, of say 100 000 years ago, were less intelligent and less capable than we are. This idea had its origins in Europe, with the discovery of Neanderthals. Although undoubtedly human, these people looked different, and their tool kit was very limited. In any case they were replaced by more 'modern' people, the Cro-Magnons. And so the very term Neanderthal came to be used as a contemptuous epithet, its popular translation being 'low-brow'.

In fact the skulls of Neanderthals, and of some comparable Asian fossils, show a brain capacity at least equal to that of average modern people. The fact that in most cases only their stone tools have survived is no guide to the full range of their tools and other artefacts. Many of those stone tools were clearly meant for use in making things from perishable materials, such as wood and bone. Even the relatively small number of tools which have survived is not necessarily related to the intelligence of the people who devised them, any more than the pre-Industrial Revolution technology of ancient Greece reflected the intelligence of Socrates, Archimedes, Euclid or Pythagoras.

Fortunately, there are groups of modern people still living in the Pacific basin who demonstrate that technology is one of the least important ingredients of succesful existence—that the things we have been clever enough to invent have been much less important to our survival than our capacity to live together in groups, to co-operate with one another. They are proof that what is in our heads is more important than what we carry in our hands.

One such group of people are the Tasaday, who live in the interior of the Philippine island of Mindanao. There are only about thirty of them, and until 1966 they lived a stone age existence in the dense mountain forests, quite unknown to the outside world. In that year an animal trapper came upon them in their remote valley, and since then their way of life has been exhaustively filmed and documented, and their isolation is gone for ever. But the anthropologists who first visited them were able to obtain a clear picture of their life before it changed, and establish the fact that the Tasaday were a remnant of the kind of people who once occupied all of the Philippines and other parts of island Southeast Asia. They give us a very good idea of those explorers who made the first water crossings from the Asian mainland to settle these islands.

The Tasaday have an extremely limited technology, in terms of manu-factured tools and other items for use in everyday life. Their most sub-stantial artefacts are stone axes—chipped and flaked pieces of flat stone, bound with strips of rattan cane to a wooden haft. They also have stone scrapers, but all their other tools are made of wood: digging sticks, bamboo knives, bows and arrows (which are used solely for hunting—the Tasaday know nothing of warfare, and in fact have no word for it). Their fire-lighting tool is a wooden drill, spun rapidly in a hole in a base-board until the heat of friction ignites small scraps of grass. They use containers made from leaves for collecting food, and bamboo tubes for short-term storage.

With the sole exception of the stone axes and scrapers, everything the Tasaday use in their daily life is perishable. Their culture will leave little trace. From a purely archaeological point of view, they would be considered extremely impoverished. And yet to understand the Tasaday is to realise that they have a full, involved and satisfying life.

The Tasaday live as a large family in a series of large rocky overhangs, under a cliff face in the forest, overlooking a clear, rocky stream. They are hunter gatherers. They hunt small deer and pigs with bows and arrows, and occasionally catch a monkey or bird in a trap, but most of their food is gathered in and close to the many streams in the forest. Frogs, tadpoles and small freshwater crabs provide animal protein, supplemented by fat wood-boring grubs found in rotting logs. With digging sticks they gather yams, which they wash and roast. The palms growing along the stream banks have edible hearts, and there are many varieties of nuts and fruits, including native bananas. Anthropologists who have lived with the Tasaday are surprised that the whole group can obtain a day's food in about two hours' foraging along the streams and in the forest.

This leaves the Tasaday with a great deal of time for purely social activities, primarily the rituals of kinship and family, and the exercise of imagination in relation to myths of the forest and the other living creatures around them. The group as a whole shows the social organisation and sense of communication and compromise which is the mark of the human animal. The only real division is that of sex; men come together to perform certain tasks and to discuss related matters, and women have parallel interests. Within the community some groups form on the basis of age — older women, young men, children. Husbands and wives, and their parents and close relatives, make up groups of another kind. So in a person's lifetime, as the individual grows up, takes a wife or husband, and has children, there are many different associations and friendships that can be made. What assured the survival of the Tasaday was their social cohesiveness, and their reliance on their capacity to work and live together rather than any range of material possessions.

This little community in the Philippines is a living clue to the kinds of people who first began to move out of Asia. And there are other clues, even in the more highly developed societies of Indonesia.

More than 2500 kilometres to the southwest, in the hills of Java, there is another example of the way in which imagination can replace technology. Here, many small communities make their living by growing rice, but the village of Cebodas is one of several that specialises in another product: bamboo, that amazing tropical plant, closely related to grass, which is in so many ways a symbol of Asia. Bamboo is one of the most extraordinary plants that exist. It grows faster than any other living organism — up to a metre a day. A bamboo plant flowers perhaps once in a hundred years, and then it dies. There are more than a thousand species of bamboo, and in its myriad uses it is the most universally useful plant known. For more than half the human race, life would be completely different without it; it touches daily existence at a thousand points, providing food, raw materials,

Bamboo is an integral part of life in Asia. Here long lengths have been made into rafts for passage down river to market in Java. *The villagers of Cebodas, near Jakarta, grow the bamboo in the forest, and their entire economy depends upon it.*

shelter, even medicine.

The people of Cebodas grow bamboo in huge groves, with individual families concentrating on particular species. It is the key to the economic life of this and other neighbouring villages. Like city dwellers mowing their lawns, the villagers cut their bamboo every week, and each week there is a market on the bank of the river. The villagers assemble the lengths of bamboo into floating rafts, each containing twenty or thirty long stems. The selling unit is a raft, and they vary in size according to the type of bamboo used. Buyers from Jakarta inspect the offerings, and make their bids. The rafts are then tied together into huge super-rafts, which are floated down the river to the outskirts of Jakarta. The men who go with them, steering them through the shallow rapids with long bamboo poles, return to the village by bus after the two-day journey. It is a trade which has its roots in the very earliest cultures of Asia.

In Jakarta the rafts are broken down, and the material goes off to be used in a host of trades. The smaller bamboos are made into baskets, children's toys, musical instruments, kites, fishing rods and lanterns. Larger bamboos are the basis of a whole furniture industry, whose products are exported all over the world. The largest stems can be seen in the spectacular scaffoldings that are used to erect high-rise buildings in Asia. These amazing webs of bamboo, which extend upwards for twenty or thirty floors, may look fragile but are actually safer than steel in these climates. Bamboo is lighter, and its flexibility enables it to yield to typhoon winds without collapsing.

The villagers of Cebodas also use their bamboo for a multitude of purposes. Today they make enough money from their sales to buy clothing, some of their food, and bricks and tiles for the framework of their houses, but virtually everything else around them is made from bamboo: floors and mats, screens, doors, beds, tables, chairs, cupboards, fences and ladders. In the kitchen, bamboo provides chopsticks, knives, ladles, scoops, cups, jugs, food containers and baskets. Most of these are made on the spot, as they are needed, and the only tool required is a small knife. A stone or shell knife once did the job just as well, so easy is bamboo to work.

And yet a village such as this, once abandoned to the jungle, would soon be reduced to the broken brick and tile foundations of houses. A future archaeologist might well wonder, in the absence of tools, what kind of an existence these people once had, and be tempted to write them off as incompetent, and certainly disadvantaged.

The misleading nature of such evidence as this can hardly be more vividly demonstrated than by applying it to the ancestors of the modern Javanese people, because we know that, regardless of their tool kit and technology, they began to expand out of Asia, to cross ocean gaps and colonise new lands, between 50 000 and 100 000 years ago.

Exactly when this movement began it is still impossible to say. But as some people certainly reached Australia and Melanesia by 50 000 years ago, that initial expansion must have begun much earlier. One of the great challenges to Pacific archaeology and anthropology is to pin down the chronology of these crucial extensions in the geographic range of modern humans in Asia.

A more difficult question to answer, perhaps, is why this process of expansion began in the first place. As we have seen, *Homo erectus* in Java stayed put for more than half a million years, and showed little physical change. In China, over the same period, there were significant advances towards the modern human form. Part of this development may well have been the growth of a more pronounced spirit of curiosity, a restlessness and desire to explore. But there may have been other, more practical, reasons. The growing ability to modify or control the environment may have resulted in local population pressures, which in turn stimulated the search for more living space. Some groups might have been forced to move by local fighting over territory.

But above all it is likely that the mainsprings of mankind's first seaborne migrations were sea level changes around the fringes of Asia, coupled with local geological and volcanic processes.

When we look at the oscillations in sea level as ice ages came and went, one important consequence immediately stands out: the vast amount of new living space that was created in shallow coastal areas as the sea level fell, only to be lost again when it rose. Along the coast of China, for example, there is a flat, offshore shelf—now submerged—that in some places is 200 kilometres wide. For periods of up to several thousand years, during cold phases, this area would have become a coastal plain. Once the rains had washed the salt out of the surface soils, plants would soon invade the virgin land, enriched by silt from the huge deposits of loess in the river valleys of China. Animals—including people—would certainly move to occupy this 'new' territory. With much the same process taking place around the coast of Southeast Asia, significant increases in population no doubt occurred. Some of these new groups may well have become entirely dependent upon these new territories.

It is not difficult to imagine the problems which arose when the climate warmed up, the ice sheets retreated, and the sea level rose again. It is known that in such circumstances the ocean can re-invade flat coastal plains at a surprising rate—up to a kilometre a year. Such a rapid loss of territory must have placed severe pressures on the populations which had settled there. Some groups may have been able to fall back inland, to be absorbed into the original populations. But for many the choice was limited: they had to learn to live with the sea, on a steadily shrinking coastal plain. Eventually, they would be forced to move elsewhere—and one way to find new living space was to use their acquired knowledge of the sea and seafaring to move along the coast and, eventually, across the ocean itself.

All this upheaval was, of course, an extremely complex process, extending over long periods. Groups of people around the shores of Asia would have been constantly on the move, finding and occupying vacant stretches of coastline. As their watercraft evolved they would have found and settled the islands they could see. After that, some must have sought land they could not see. And so it seems likely that the many oscillations of sea level over the past 100 000 years or so produced a series of migratory pulses out of Asia, with people moving each time the sea began to rise.

Other influences on those early movements of population were undoubtedly the violent geological processes along the so-called 'ring of fire' on the western margins of the Pacific. From examples which occurred in historical times we know the effects of such events on coastal peoples. The most spectacular and best documented example was the eruption of Krakatau in 1883. This volcanic island in the Sunda Straits, between Java and Sumatra, literally blew up, with an explosion that was heard 5000 kilometres away. It caused *tsunami*, or tidal waves, that killed more than 30 000 people in the Indonesian island chain. Thousands of coastal communities had to move, and make new lives elsewhere. An earlier but less well-known volcanic explosion on the island of Sumbawa, east of Bali, is believed to have killed more than 90 000 people in Southeast Asia, either

The bamboo rafts still used by fishermen on the south coast of China are perhaps clues to the watercraft which carried the first people out of Asia into the islands of Southeast Asia, more than 50 000 years ago. *Some larger rafts, used along this coast until recently, could carry several tonnes of cargo, and stay at sea for weeks.*

directly or as the result of famines brought about by devastating *tsunami* and the destruction of food-producing areas. Undoubtedly such events have occurred many times in the past, with incalculable effects on coastal populations in Asia.

Whatever the complex of reasons involved, it is certain that groups of people from mainland Asia began making ocean crossings well before 50 000 years ago—which leads to perhaps the most intriguing question of all: how did they do it?

There are suggestions, of course, that these first sea voyages must have been accidental—that people were blown out to sea on floating logs, or something similar—because first, they could not have known that their destinations even existed, and second, they were not sufficiently advanced to have developed watercraft able to make such voyages.

We firmly believe, however, that while the discovery of unknown islands could not be said to have been planned, the people who made those landfalls succeeded because they were already using surprisingly capable watercraft. Just as the Polynesians, much later, were able to make their astonishing explorations of more than 4000 kilometres across open ocean to places like Hawaii because they had already developed their canoes in trade between the nearer islands, so those early Asians discovered the Philippines, Melanesia and Australia because they had simple but seaworthy watercraft that could easily cross the short sea gaps involved.

What kind of watercraft could they have used, more than 50 000 years ago? Whatever they were, they must have been made from perishable materials. It is therefore very unlikely that we will ever find any trace of them in the archaeological sites that are available to us today. Not only do objects made of plant or animal materials deteriorate rapidly in warm, humid climates, but the coastlines where those first landings took place have been covered by the most recent rise in sea level. However, after considering the areas where the first seafarers must have lived, and the materials which were available to them, we found strong clues to the nature of those early watercraft along the coast and rivers of China—a living museum of the history of water transport.

In coastal towns and villages there, fishermen still use a form of watercraft which, in materials, design and function, could easily be 50 000 years old. The skills needed to build it would certainly have been possessed by the people of those times, and we have proved, by experiment, that it is quite capable of making crossings of the open ocean.

This remarkable watercraft is a simple bamboo raft, made of up to eleven stout lengths of bamboo lashed side by side, braced by thinner bamboo crosspieces. The lashings are strips of the outer skin of the bamboo, which contains long, thin fibres as strong as steel wire. The result is a platform between one and two metres wide, and up to about four metres long, tapered at the front. The long bamboos are slightly turned up at the bow, to better ride the waves, and there is a slot built into the stern

Bamboo rafts have been used on rivers and lakes in China for thousands of years. These are on the Li River at Guilin. *The cormorants are used by their owner to catch fish at night.*

to take a steering oar. Amidships, the raft may have a step to take a mast and sail; otherwise it is rowed by the fisherman, who stands up facing forward and pushes on his oars, in the Asian fashion.

A raft like this is flexible yet strong, and easily withstands the pounding of heavy waves. It is quite unsinkable because of the buoyancy of the chambers in the bamboo, and if it is taken out of the water occasionally to kill marine borers it will last for years. Fishermen go to sea on these rafts for days at a time, fishing in groups, often well out of sight of land; a powered mother ship comes round to collect their catch. Similar rafts are used on the Li River near Guilin, famous for its towering limestone pillars. Local people use the rafts to cross the river, and to fish with nets. Professional fishermen use them for cormorant fishing at night, attracting fish within reach of their tethered birds with a lantern. These rafts are poled or paddled along.

Until recent times larger versions of the bamboo raft were in use along the Chinese coast for carrying cargoes. Some were as much as seven metres long, and could carry several tonnes of cargo. In fact it seems likely that the bamboo raft was the ancestor of that peculiarly Chinese craft, the junk. This interesting vessel, unlike most Western craft, has no keel, but a smoothly rounded bottom. Its hull is divided by a number of bulkheads, creating watertight compartments which make the junk virtually unsinkable; in this it closely resembles a bamboo. (This principle of ship-building was only adopted in the West in the 1840s, with Isambard Kingdom Brunel's revolutionary iron ship *Great Britain*.) What could well be a transitional stage can still be seen on the coast of China today, where some boatmen have built up the sides of their rafts with extra bamboos. This gives them protection from the waves, and enables them to carry more. The general shape of such a bamboo boat is very similar to that of the junk.

The first bamboo rafts were probably built once early Asian people had begun to cross rivers and bays on floating tree trunks or mats of vegetation. Next they would have learned to propel them, first with poles and then with paddles and oars, and to steer them. And anyone who stood up on such a raft in a breeze would rapidly grasp the principle of the sail.

It is for all these reasons—the simple technology of the bamboo raft, the availability of its materials, its seaworthiness, and its existence for an unknown period on the coast of China—that we believe that something very like this watercraft was the key to those first oceanic migrations out of Asia. This belief was reinforced by personal experience, which one of us (Alan Thorne) describes as follows:

'Because of my interest in the biological origins of the first Australians, and the ways in which people might have begun the colonisation of island Southeast Asia, I began to take an interest in bamboo rafts after my first visit to China in 1973. By the early 1980s I had seen quite a variety of rafts, sampans and junks in action, and had even made a few sailing trips with fishermen on their bamboo rafts along the coast near Shantou, in Guangdong province. But to get a better perspective on the question of whether bamboo rafts could have been the kind of vessels that first reached Australia I decided to have a large one built in Indonesia, and try it out.

I went to Cebodas, the bamboo-growing village near Jakarta, bought a raft and a large number of spare bamboos, and persuaded the headman and two of the villagers to build me a sea-going raft, and do some test sailing. The experiment was carried out among a group of atolls called Pulau Seribu—the Thousand Islands—in the Java Sea. The raft we built was about fifteen metres long, two metres wide at the stern, and tapering to a graceful, slightly upturned prow. It had a short bamboo mast, and a sail that we stitched together from a couple of woven sleeping mats that I bought in the market. Two long steering oars were made from bamboo poles, with split sections lashed at right angles. All lashings were either strips of bamboo skin or woven vines and bark. The whole thing took only a couple of hours to make.

The villagers from Cebodas took a little while to get used to handling our craft on the sea, but even on that first day the four of us took it from one island to the next, across deep, open water. The raft was surprisingly easy to steer and control. Once out of the wind shadow of the island, with a stiff breeze behind us, we moved freely along at four or five knots. The long bamboos were very flexible, and the raft rode easily across waves and troughs, giving us a most comfortable ride. The deck was constantly awash, but a small superstructure would have enabled us to carry food

The distinctive design of the junk may have evolved from the bamboo raft. It has a round bottom and watertight compartments—just like a piece of bamboo. *Junks like this are still a common sight on the rivers and coastline of China.*

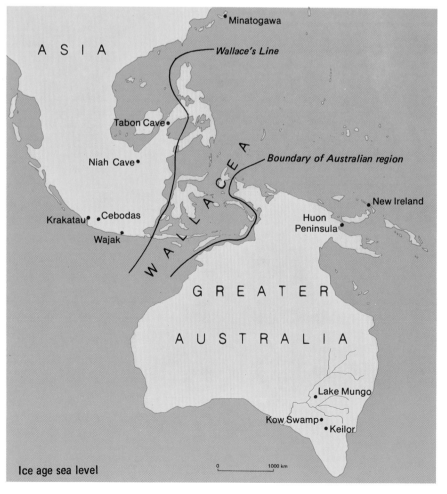

The first new territories to be occupied by people moving out of Asia were the many islands of the Philippines and Indonesia. *This is part of the Indonesian island chain, east of Bali.*

and other requirements for a serious voyage. Fresh water and food could also have been stored in some of the hollow compartments in the bamboo. I have no doubt that with proper preparations, and the right winds and weather, we could have island hopped for a considerable distance, and even crossed quite large stretches of open ocean.'

Multiply this experience along the Indonesian island chain, with the monsoon winds blowing steadily from the northwest, and it is clear that even such simple watercraft could have reached the coast of Greater Australia—especially at times of lower sea levels. The water gaps between the Asian mainland and Australia never completely disappeared, but for long periods during the ice ages there was no Malacca Strait between Malaysia and Sumatra, no Torres Strait between New Guinea and Australia, and no Bass Strait between Australia and Tasmania.

To deliberately set out to cross that last water gap between the Indonesian islands and the northwest coast of Australia, would seem, at first sight, to be a colossal gamble. The distance was probably too great for people to be able to see Australia, but it has been suggested that large bushfires on the Australian mainland, started by lightning, might have produced columns of smoke visible by day, or a glare on the horizon visible by night. The southeast trade winds, which blow from Australia towards Indonesia, might have carried floating debris, suggesting the presence of land somewhere in that direction. Any of these signs might have been enough to encourage intrepid voyagers to set out on their rafts. And then there is always the question of chance.

Data from the raft 'voyage' were used to set up a computer simulation experiment at the Australian National University in Canberra. A computer was programmed to imagine that a series of rafts, without sails, were simply blown away from the coast of Timor, near the eastern end of the Indonesian island chain, during the monsoon that blows towards Australia from November to March. The known pattern of ocean currents and variations in wind strength and direction were fed in, and the random movements of the rafts were plotted. Remarkably, nearly every raft finished up on the Australian coast sooner or later, most within a week or ten days. And these results were obtained at today's sea level; when the computer was reprogrammed for a typically low ice age level, it appeared literally impossible for such a raft to miss Australia.

So it seems quite likely that the first actual landing in Australia could have been made by small groups of people—perhaps fishermen, perhaps a few families trading along the islands to the northwest, or looking for a new home—who were simply blown off course and saw new land on the horizon.

Of course, the discovery of Australia was the end of a whole sequence of much shorter voyages between the Asian mainland and offshore islands, and then on from one island group to another, over many thousands of years. One line of approach to Australia was probably the one just described, southeast from Indonesia, or perhaps due east into New Guinea (which for long periods was part of Greater Australia). But there was another line of approach, from China in the north down through the Philippines. As we shall see, there is evidence to suggest that people arrived by both routes, from quite different cultural backgrounds.

Whichever way they came, we now know for certain that the human settlement of Australia began at least 40 000 years ago, and there are indications that it could have begun closer to 100 000 years ago. This possibility depends upon the fact that the earliest archaeological evidence of human occupation in Australia is found, not in the north, but in the southern half of the continent. People were present 40 000 years ago at archaeological sites found at Keilor, on the outskirts of Melbourne. At the now dried-up Lake Mungo in western New South Wales there are firmly dated human remains from 38 000 years ago. Stone tools of the same age have been found on the outskirts of Perth, in Western Australia. Since it seems very unlikely that these few sites are the oldest that exist, their dates must be considered a minimum when assessing the antiquity of the human presence in Australia.

Another question is how long it took people to reach the southern half of the continent, after the initial landings on the north coast—especially as that coast was, for much of the time in question, much further to the north, in New Guinea. Large stone axes have been found on the Huon Peninsula on the north coast of New Guinea, cemented by volcanic ash into one of the uplifted fossil coral reefs mentioned earlier. Dating of this fossil coral shows that people were on that coast 40 000 years ago. Another discovery of tools 30 000 years old in New Ireland, off the New Guinea coast, shows

Kow Swamp, in Victoria, was an early centre of population in Australia, because of the rich food resources in and around the lake. *These large freshwater mussels were a favourite food of the first Australians.*

Winds sweep the dunes along the dried-up shores of Lake Mungo, in western New South Wales, exposing the shells of mussels consumed during the ice age. *Lake Mungo is one of the Willandra Lakes, which at that time were full of water and alive with birds, fish and shellfish.*

that the early sea-going explorers had moved even further east by that time.

All this evidence means that the first colonists had begun to arrive in Australia by 50 000 years ago, and by about 20 000 years ago they had spread into every corner of the continent. Given the time it must have taken for the waves of human migration to reach and cross the island groups of Southeast Asia, the original movements out of mainland Asia must have started well before that date.

The extreme limit of that first great push into the Pacific basin, and the most important new territory to be occupied by humans, was the continent of Greater Australia—an area as large as all the lands and islands of Southeast Asia combined, even at the time of maximum ice age exposure. In the next chapter we will see just how the first colonists adapted to their new environments, and how their descendants raised their technologies, skills and cultures to perhaps the highest levels achieved by hunter gatherers anywhere in the Pacific basin.

But before we leave the subject of that initial settlement, there is the intriguing question of where *exactly* the first Australians came from. Their physical characteristics, as defined by the earliest human remains, must be a guide to the origins of those migrants—and the Australian human fossil evidence, especially that discovered in the last twenty-five years, is rich in clues.

One of the most significant sites is Kow Swamp in northern Victoria, near the Murray River. Kow Swamp is a large freshwater lake, and between 10 000 and 15 000 years ago people lived round its margins. They hunted kangaroos in the fringing woodland, caught Murray cod, golden perch and tortoises in the lake, and dug up large freshwater mussels from the mud flats. They also buried their dead with some ceremony in the soft silts near the lake shore. Some of the graves were decorated with red ochre, and shells and stone tools were placed with the bodies.

In the 1920s, when the Murray River Irrigation Project was being developed, earth-moving work disturbed some of the Kow Swamp burials, but their nature was not recognised. It was not until the early 1970s that a chance inspection of some human bones in the Museum of Victoria in Melbourne by Alan Thorne, and a search for their origin, resulted in the excavation of the Kow Swamp site, which he describes:

'I recovered the remains of more than forty men, women and children. All of them had been disturbed by the earlier earth-moving. What immediately made the most striking impression was the anatomy of the skeletons. All the people were quite robust, taking age and sex into account. Their skulls were thick, ruggedly constructed, and had some features which are not seen in modern aboriginal Australians. The faces were very large, as were the teeth, and projected markedly forward. The brow ridges were thick, and in some cases form a complete buttress over the eye sockets. Foreheads were flat, sloping back to a low crown on the cranium. It was clear that these people shared a complex of features that I had seen outside Australia, in Indonesia. Similar remains have also been recovered at Cossack in Western Australia and at Talgai in Queensland.

The anatomical characteristics of aboriginal Australians had suggested for a long time that their ancestors came from Indonesia. Just as *Homo erectus* in Java had contributed directly to the early *Homo sapiens* people in Ngandong, so the Ngandong feaures could be recognised in the early Australians. The Kow Swamp and other human fossil remains were strong evidence for the suggestion that Indonesia had made a substantial contribution to the Aboriginal population. But anthropologists had always suspected that this was not the full story, and variations in other Aboriginal skeletal remains prove that the ancestry of the first Australians was not wholly derived from Indonesia.'

The discovery in 1968 of human remains at Lake Mungo, one of the now dry Willandra Lakes in western New South Wales, revealed another side to the story. The first find, eroding out of a wind-scoured sand dune, was a small block of harder sand containing a few broken bones, some blackened by fire. When painstakingly cleaned and reconstructed, the bones were found to be the cremated remains of a young woman. They were dated to about 25 000 years ago, which makes this the oldest known cremation anywhere in the world. [2]

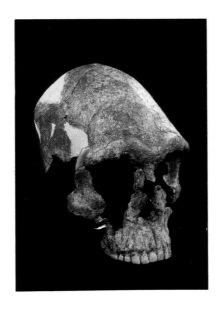

One of the skulls excavated by Alan Thorne at Kow Swamp in the 1970s. *This cranium, with its prominent brow ridges, shows a clear link with Java Man—one of the two ancestral roots of the Australian people.*

More interesting, as far as our story is concerned, is the nature of the bones themselves. They are remarkably gracile (delicate), and in fact the skull is so thin that at first the fragments were thought to be pieces of an emu egg shell. In contrast to the Kow Swamp skulls, the Lake Mungo woman has a skull that is not only delicate but very modern in shape, with an upright and rounded forehead and crown, and teeth that do not project forward. In 1974 the burial site of an adult man was found a few hundred metres from the cremation site, with red ochre scattered across the skeleton. This burial was older, dating to about 30 000 years ago, and the features of the bones confirmed those found in the cremation site. The only other skeletons of this age, showing these skeletal features, have been found in China.

A detailed comparison has shown beyond doubt that the origins of these more modern people at Lake Mungo must lie in China. The remains from the Upper Cave at the Peking Man site at Zhoukoudian (Dragon Bone Hill), as well as those from Ziyang and Liukiang (described in the previous chapter), are remarkably similar to those found at Keilor and Lake Mungo. Very similar skeletal remains have also been found at a number of sites lying between China and Australia: Minatogawa on Okinawa, Tabon Cave on Palawan in the Philippines, Niah Cave in Sarawak and Wajak in Java.

This chain of sites shows that during the ice ages people from the Asian mainland pushed south into the islands of Southeast Asia. The Lake Mungo skeletons show that some of them, or their descendants, made landfall in Australia. In the course of this expansion these largely China-based colonists replaced the more robust people of Indonesia, some of whom had already begun to enter the New Guinea-Australia area. The Asian newcomers had lighter skins and more delicate features, and we see their descendants in the modern brown-skinned Indonesians. The original people from Indonesia provided the basis for the dark people of New Guinea and other parts of Melanesia. In Australia these two strains of people—as represented by the remains at Kow Swamp and Lake Mungo— came together, and eventually produced the modern Australian Aboriginal.

Of course the development of the Australian population was hardly as simple or as straightforward as that. It was an extraordinarily complex process, involving both physical and cultural integration, spread over perhaps 50 000 years. It did not consist of two boatloads of different people who later intermingled, but an endless series of arrivals at different places and from different directions, joining others who had come before and scattered across the landscape. We have no way of knowing exactly when it began, who came first, or whether there were any large gaps in the sequence of arrivals.

It is best to think of it, once it had started, as a steady flow of adventurous, hardy, adaptable people out of Asia, through the islands, and down into Australia. And they only stopped in Tasmania because, on this side of the Pacific, that frigid, glaciated, mountainous corner of the continent was as far as they could go.

The skull of one of the people who lived at Lake Mungo 30 000 years ago, partly exposed by wind erosion. *This was the way the first human remains were discovered here in 1968.*

The same skeleton, during excavation. Red ochre was sprinkled on the bones of this man before burial. *The remains of 140 individuals have now been found around Lake Mungo, one of Australia's most important archaeological sites.*

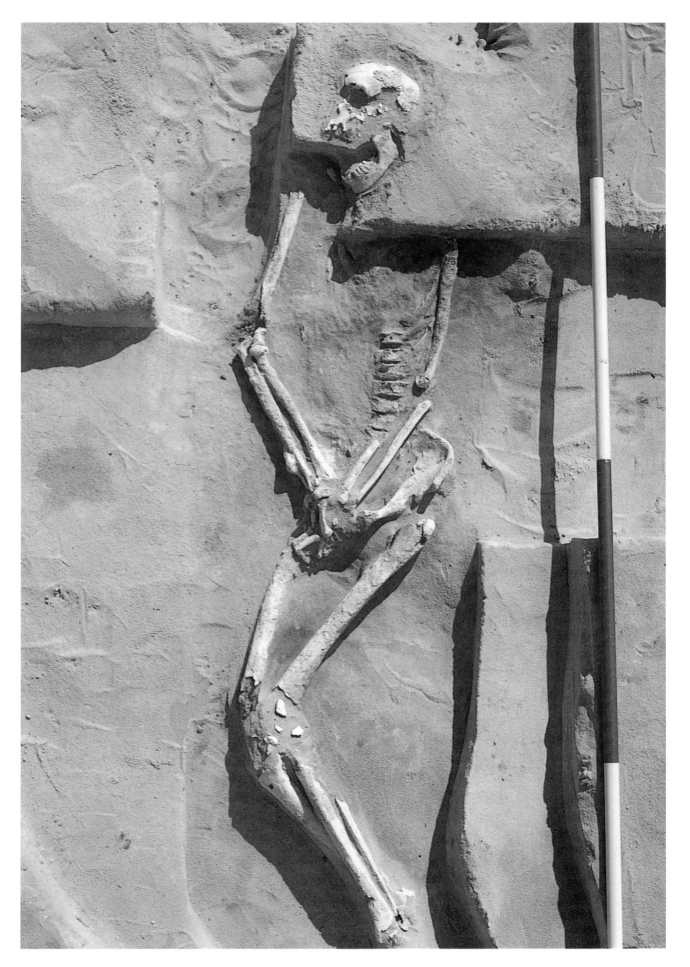

3.

Hunters and Gatherers

What the physical conditions were like on the Australian continent when the first people arrived is still largely a matter of speculation, partly because of the uncertainty about when that event took place. But if we take the period in which we are fairly sure it happened—betweeen 50000 and 100000 years ago—we can make some assumptions concerning the landscape, the climate, the vegetation and the kinds of animals that were living in Australia at the time.

As discussed in the previous chapter, that whole period saw enlargements and contractions of the polar ice caps, with subsequent changes in world sea levels and climatic conditions. The most severe glaciations were in the Northern Hemisphere, but at those times ice age conditions also applied in Australia. Over the long period from 50000 until about 10000 years ago the details of sea levels become clearer and we can describe more accurately what the continent looked like to its early inhabitants.

In the north a broad plain, dotted with hills and probably well grassed, joined Australia to New Guinea. The area now occupied by the Great Barrier Reef was also a flat plain, sometimes extending to the edge of the continental shelf. Fairly narrow coastal plains extended down the eastern and western coasts, with hills where the present offshore islands are. Sydney Harbour was a deep valley cut into a sandstone escarpment by rivers, and the coastline was many kilometres further east. Tasmania was linked to the mainland by land, with a ridge of low mountains to the east. There was also a wide, very flat plain filling the Great Australian Bight— an extension, although at a lower level, of the Nullarbor Plain.

Perhaps the most unusual features of the continent, compared to what we are used to, were the vast freshwater lakes. The largest was Lake Carpentaria, which occupied most of what is now the Gulf of Carpentaria. It was some 500 kilometres long and 250 kilometres wide. The next largest was Lake Bass, 250 kilometres long, which lay where Bass Strait now is. Its existence meant that the links between the mainland and Tasmania were

In the more than 50000 years that they have occupied the continent, the first Australians have left enduring marks of their presence here. *Rock engraving in Mootwingee National Park, in western New South Wales.*

actually two quite narrow land corridors, on either side of the lake.[1]

Both these lakes, although fed by large rivers, were probably brackish, but they would have supported rich fish populations and huge colonies of birds. Each would certainly have become a major focus for early aboriginal people. The same environment, rich in aquatic and bird life, was provided by the chain of large freshwater lakes which extended across the centre of the continent, from South Australia to Lake Carpentaria. As we have already seen, the Willandra Lakes, including Lake Mungo, became early centres of human population.

Overall, the climate across the continent was considerably cooler and drier than it is today, and the huge arid region in the centre and western part of the continent even more desiccated. This was the time when fierce winds built the thousands of sand dunes which today extend like a petrified ocean across large areas of the inland. In the higher parts of the continent, in the southeast ranges, conditions were much colder than they

are today, and the winter snowfields more extensive.

Elsewhere, however, the various types of wet and dry eucalypt forest, open woodland, heathland and grassland may not have been much different from what they are now. These plant communities had already adjusted to the steady cooling and drying of the Australian continent which followed the end of the warm, moist Tertiary period, some 1.6 million years ago. The rainforest, which once extended right across the interior, had retreated to its coastal refuges, extending from north Queensland to Tasmania. The drought-adapted and fire-resistant eucalypts (450 species) had taken over as the dominant trees, and the hardy acacias (900 species) made up the forest understoreys and mulga scrub of the arid interior.

One environment, the coastal belt across the tropical north, stood out from the remainder of Australia then (as it still does) because of its uncharacteristic vegetation. This it acquired after the crustal plate carrying the Australian continent drifted up against Indonesia, about four million years ago. Many plants from the Indo-Malaysian islands took the opportunity to invade the newly available territory (although the reverse migration of Australian species did not seem to happen).

When the first people from Southeast Asia landed on the coast of Greater Australia they would, therefore, have felt immediately at home in the mangrove swamps and pandanus groves along the shoreline, and in the rainforest on the lowlands. They would have been able to seek out familiar plant foods: fruits, nuts, yams, succulent shoots and bulbs.

These first Australians would naturally have followed their traditional food preferences and methods of preparing food, rather than experimenting with strange new plants. In fact these habits persisted in northern Australia right down to the present; many edible native plants were ignored in favour of those which are also found in Southeast Asia. Jack Golson, of the Australian National University, has listed the food plants used by modern people in Arnhem Land. Out of sixty-five preferred species, no fewer than twenty-eight are also well known as food plants in Indonesia, Malaysia and the Philippines, and another twenty-seven are related to food plants used there.[2]

Some Asian plant foods, such as mangrove seeds, palm nuts and some yams, contain poisonous alkaloids and require various forms of heat treatment and leaching with water to remove their toxins. The first colonists obviously brought this knowledge with them, but as they spread out from their original beach-heads they would have found fewer plants that they knew, and more that were unfamiliar. The experimentation with strange plants which followed may have taken thousands of years—and cost many lives—but in the end the aboriginal pharmacopoeia expanded to include hundreds if not thousands of food plants and medicines.

Among the drugs which the early Australians discovered and used were narcotics. They obtained a popular and powerful stimulant by chewing the leaves and twigs of a shrub, *Duboisia hopwoodii*, known as pituri. They used branches of pituri, as well as a number of other plants, including a eucalypt and a mangrove, to poison waterholes and lagoons in order to

The first people to land on the northern coast of Australia would have found an environment very like the one they had left in Southeast Asia. *In the Gulf country of Queensland, mangrove-lined rivers wind across tidal salt flats to the sea.*

catch eels and other fish. Even large animals such as emus and kangaroos which drank the water became stupefied and were easily speared. Other plant materials were used to treat sickness and promote the healing of wounds, in ways which foreshadowed the modern range of fungal anti-biotics, such as penicillin.

In fact the pragmatic botanical knowledge of hunter gatherers, and of Australians in particular, is one of the great intellectual accomplishments of the Pacific peoples, although one of the last to be acknowledged. It was certainly not recognised until too late by Captain Cook's men, who ate some of the seeds that were popular with the local people in north Queensland; they even gave some to the pigs on board *Endeavour*. But because the seeds had not been treated, the sailors got sick and the pigs died.[3]

Undoubtedly the biggest surprises for the first arrivals in Australia were to be found in the animal kingdom—although not, perhaps, in the shallow

Ice age sea level

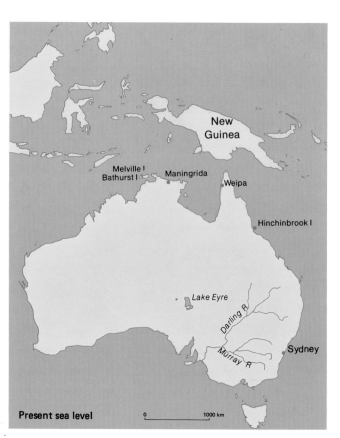

Present sea level

coastal waters and river estuaries, which contained many of the same fish, crustaceans, turtles, dugongs and crocodiles that they had caught in the islands of Southeast Asia. But as they began to move south they must have begun to glimpse shapes they did not recognise, bizarre animals that they could not catch.

Those animals would have been almost as strange to us, because they included many giant creatures that are now extinct, and known only from fossils. The largest were the diprotodons, a group of bulky, wombat-like grazing marsupials which were common on the grasslands of central Australia. There were several species, varying in size, and the largest was

a

b

In Australia the hunters found a range of animals whose mobility made them harder to hunt than the inhabitants of the Southeast Asian rainforests, so they increased the efficiency of their spears with the woomera. (a): *The woomera is a short stick with a hook at the tip, which fits over the end of the spear. (b): To throw the spear, the hunter grips the woomera and the spear shaft together. On the throw he holds the end of the woomera in his fingers, so that it acts like an extension of his arm to give increased leverage and thus greater impetus to the spear.*

the biggest marsupial that ever lived. There were also giant kangaroos and huge emu-like birds. At the top of the food chain were two large marsupial predators: *Thylacoleo* (a marsupial 'lion'), which had oversized shearing teeth for tearing lumps of meat from its quarry, and *Thylacine* (the Tasmanian 'tiger'), a wolf-like animal which ran down its prey by persistence.

In learning to hunt all these,[4] as well as smaller kangaroos and wallabies, possums, gliders, wombats, bandicoots, platypuses, echidnas and native cats, the first colonists would have encountered another new factor: the nocturnal habits of most Australian mammals. They are active at night, but by day they sleep or rest, and are hard to find. Thus the hunters developed their renowned skills at tracking animals and digging them out of their burrows and hollow trees and logs.

One indication of the hunting techniques developed by the early Australians was their specialisation with certain weapons. They became particularly proficient with the spear, which enabled them to kill animals at a considerable range in open country, or when their quarry was drinking at a waterhole. The effectiveness of the spear was increased by the use of the woomera. This is a flattened wooden stick with a hook which fits over the butt of the spear; the other end is held in the hand which grasps the spear shaft. On the throw the woomera acts as an extension of the arm, giving extra leverage and therefore greater velocity and distance.

Another hunting tool was the boomerang. This aerodynamically sophisticated device (which is known from other parts of the world, including Egypt) may have developed from a simple throwing stick, used to knock animals out of trees. It is not hard to imagine this evolving into a flattened, curved throwing stick for bringing down birds taking off from a swamp. The refinement of the return boomerang may have been simply a convenience, to save the effort of retrieving the missile.

One weapon the Australians do not appear to have used is the bow and arrow. This was widely known in Southeast Asia, but if it ever reached Australia it was subsequently abandoned. The bow and arrow is still used in New Guinea, and was known to Australian hunters. There is no certain explanation for its rejection, but it may be that the arrow, while deadly against monkeys and birds in the close quarters of the jungle, was less effective against the larger animals in the more open spaces of Australia. There, the spear may have proved more effective.

'For humans', as the ecologist L. J. Webb has pointed out, 'botany began anew in Australia', and the same revolution was demanded of the whole practice of hunting and gathering. It was in the ingenious ways they solved those challenges that the Australians refined the life-style of the hunter gatherer to a degree not seen anywhere else in the Pacific basin.

One measure of the Australians' originality and versatility can be found in the wide range of their tools, weapons and utensils. Very few of these were made of stone; most were fashioned from an amazing variety of perishable materials. In many long-vanished cultures such artefacts simply disappeared with time, leaving only stone tools to provide a picture—often misleading—of apparent impoverishment. Fortunately, the persistence of aboriginal culture right down to the present day enables us to see virtually the full range of its material expression.

Stone is, of course, used in many ways—to make spear points, sharp edges for cutting meat, scrapers for cleaning skins. Suitable stones are also flaked to make axes, and in the distant past some were sharpened by having their edge ground down to a smooth surface. A number of these unusual edge-ground axes have been found in Kakadu National Park in Arnhem Land, and dated to more than 20000 years ago.[5] The only other place in the world where axes of this kind and of comparable age have been found is Japan, although some slightly later ones have turned up in Southeast Asia. All of them are at least 10000 years older than any similar axes from Africa or Europe. Archaeologists are at a loss to explain the apparent affinity between the axe-grinders of Arnhem Land and Japan, all that long ago, but we believe it is related to the early routes of migration out of China into the Pacific basin.

For Australians, the crucial stone implement is the so-called core tool. The tool-maker 'sees' this locked up within a rock, and reveals it by knocking off all the flakes of stone that surround it—just as a classical Greek or Roman sculptor discovered a statue within a block of marble. Some large core tools are really heavy planes, used to shave down the shaft of a spear, or to gouge out the basic shape of a dish or shield. One of the most useful tools consists of a quite small, sharp-edged flint or piece of quartz fixed to the handle of the woomera with rock-hard spinifex gum, made by melting together small particles of gum from the leaves of spinifex grass. This chisel-like, multipurpose blade is used for cutting skin and sinew, and fine planing and sharpening of wood and bone.

The great bulk of aboriginal artefacts are, however, made of plant materials. Wood from a variety of trees is made into spears, boomerangs, clubs, shields, digging sticks, dishes, animal traps, house frames and dugout canoes, as well as many items of ritual significance, including funeral posts, sacred boards and masks. Twigs and grasses are woven into a huge variety of baskets, fish and bird nets, strings for binding and making necklaces, and mats for sleeping or wrapping the dead. Bark stripped from trees makes a roof, a painting surface, or a canoe. Plant juices make dyes, flavourings and medicines.

Animal products are put to innumerable uses. The skins of kangaroos, wallabies, possums, koalas and platypuses make rugs and cloaks. Sinews make strong lashings, and wallaby and bird bones are transformed into scrapers, needles, toggles for fastening cloaks, and barbs on fishing spears. Shells are used intact for carrying water or baling out a canoe. Others are cut into fish hooks, or engraved as jewellery. Even human hair is twisted into cords, and made into carrying bags and nets.

Most of these utensils and implements are discarded and reduced to dust within a few years of being made. For most Australians, the availability of raw materials and the ease with which they can be transformed into something useful makes durability superfluous and preservation un-

A reliable source of food for coastal hunter gatherers are the eggs of the green turtle, buried in the sandy beaches. *The people of Bathurst Island, who still follow a largely traditional life, are the darkest of all Australians.*

Another highly prized food in the north is a fat, juicy 'worm' dug out of dead mangrove trees. *Although they look like worms, these are actually a species of mollusc which burrows into wood instead of making a shell.*

necessary. A large range of labour-intensive ceremonial objects, such as carved funeral posts, sand paintings on the ground and the complex decorations and costumes worn in ceremonies, take a great deal of time and care to create—but they are made for a single event, such as a funeral, and then discarded. The aboriginal Australians put no store in antiquity as such. Being by tradition nomadic, they have no wish to be encumbered with unnecessary possessions which have fulfilled their purpose.

The full story of the Australian people is beyond the scope of this book, and in this chapter we can give only an outline of the way in which they occupied a continent as large as Europe or the continental United States.

Whatever the date of the original arrival, we know from archaeological evidence that groups of people had explored all the available territory by about 20 000 years ago—just about the middle of the last ice age, when the sea level had reached its lowest ever. There is a date of 23 000 years for a habitation site in a cave at Cave Bay, in northern Tasmania, and another of 20 000 years even further south, in Kutakina Cave on the Franklin River. People had also begun to move into the desert quite early; a recent excavation in a cave at Puritjarra, to the north of Ayers Rock, shows that people were living there 23 000 years ago. Around 20 000 years ago some groups were camping at Birrigai and Cloggs Cave, in the uplands of southeastern Australia; at this time there would have been glaciers in the higher valleys.

To obtain some idea of the diversity which developed among the Australians as they spread out to occupy the continent, we will look, as we did in our film series, at some of the widely varied environments which exist today, and use these as windows into the past, to a time when they supported large populations. Fortunately, many aboriginal tribal societies have survived, with their culture, and can provide living testimony to the complex lives of hunter gatherers in Australia over the past 50 000 years or more.

We begin where the human history of Australia began—on the north coast, where the tropical sun beats down on blue-green sea and dazzling white sand, and the warm, scented breeze rustles the leaves of the pandanus palms and whispers through the drooping casuarinas. In the dry season the days and nights are calm and clear, and even during the monsoon the clouds rarely roll in until afternoon. It is an environment that murmurs with life, and the pulse of growth and proliferation. So it is not surprising that the first Australians settled down here, and their descendants have continued to live here for the past 50 000 years or more. Theirs is a continuity and a link with the human past that is to be found today in few places, anywhere in the world.

In the Arafura Sea north of Darwin, forty kilometres off the coast, lie Bathurst and Melville Islands, separated by a narrow passage. They are the home of the Tiwi people. Because of their isolation, and hostility from the mainland tribes, the Tiwi developed a distinctive culture, and they have managed to preserve many aspects of it. To watch the daily life of the Tiwi is like watching the activities of the first Australians who lived on this coastline.

The Tiwi, perhaps because of their northerly location, are deeply black—the darkest of all Australians. They tend to be tall and well-built and have curly or wavy hair. Their lives revolve around the extended family. A man may have a number of wives—a senior man as many as ten or twelve. With children, this group forms an effective social and economic unit, particularly in the daily search for subsistence. Traditionally, the men and women went out separately, the men to hunt and the women and children to gather, but today these activities are more communal.

The two islands are well stocked with wallabies, large goannas and possums. These are hunted from time to time—especially when the seas are rough—but the marine foods are more highly prized, and usually easier to obtain. From their dugout canoes the men spear fish in the shallow waters inside the reef, and occasionally corner a large green turtle, or find a dugong browsing on the beds of sea grass. Both these large animals are protected, but aboriginal Australians are permitted to take them for their immediate use. When cut up and distributed either one provides protein for a large group. Such traditional exploitation of these resources offers no threat to their survival.

The turtles provide another welcome source of food when they come ashore to lay their eggs. They do this by scooping out a deep hole with their flippers, depositing up to a hundred golfball-sized, leathery-shelled eggs, and filling in the hole. They lay their eggs under cover of darkness, but the tracks they leave on their way back to the ocean provide a simple guide to the nest site. The men locate the eggs by probing gently in the sand with a thin stick, and it takes only a few minutes to dig them out. The problem of how to carry fifty or sixty eggs is easily solved by wrapping them in a bundle of grass and tying it up with beach vines.

Turtle eggs are a useful fall-back for the hunters; sometimes they will be the only prize they bring back to camp. Despite all their skills, and the time they spend tracking and stalking wallabies, or paddling the lagoons looking for fish, turtles or dugongs, the men often return empty-handed. But while the men have been out hunting, the women and children have been foraging—and they always come back with food. This is in fact the reality of the hunting and gathering way of life—it is gathering, not hunting, that consistently provides the daily food for the group. There has long been this image of man the hunter, coming back to his family with a kangaroo over his shoulder. But while these large game are highly prized and keenly sought after, it is the small animals and plant foods caught and gathered by hand that supply the daily bread. And in that sense the Tiwi women are the breadwinners, because they do most of the gathering.

There are many places to forage for food on Bathurst and Melville Islands, but among the most productive—although least promising, to outside eyes—are the mangroves. These remarkable trees, which can tolerate salt water, grow closely together in muddy bays and estuaries.

Rainforest

Sclerophyll
forest

Woodland

Grassland

Shrubland

Desert

Beneath their tangled arches of roots, alternately submerged and exposed by the tide, the Tiwi women find a delectable range of foods. There are monster mud crabs, with claws that can break an unwary finger. The women carefully drag them out with a stick and then expertly snap off the powerful pincers. Buried in the mud there are large numbers of fat cockles, easily located with probing toes; a kilogram or two can be collected in half an hour. There are snails to be picked off the stilt roots, nests with eggs, and the occasional tree snake. Even the rotting mangrove trunks have something to offer: a thick, juicy 'worm', up to twenty centimetres long, which is actually a soft-bodied mollusc.

Hunting and gathering is usually done in the morning, and is over before the sun gets too hot. In the heat of the day the men, women and children come together again, beneath shady trees near the beach, to eat their main meal of the day. A typical Tiwi midday meal may include whole fish, strips of turtle meat, turtle eggs, or a pile of mangrove shellfish.

Australians have never used clay pots or any other containers for cooking, so nothing is ever boiled or stewed. Food of all kinds, from kangaroos to shellfish, is grilled in open wood fires, or baked beneath a covering of hot ashes and sand. Shellfish are traditionally arranged in a neat square, each with its powerful hinge uppermost, and a small fire of twigs and leaves built on top. It takes only a few minutes to cook the succulent contents. The women rake the bivalves out of the ashes with a stick and open them with a single, practised tap.

Another rich and seemingly inexhaustible source of food for the Tiwi women are the beach flats and foreshores exposed at low tide. Here they sit, and with a small digging stick gather piles of pipis and other bivalves, which they pack into dillybags woven from coloured grasses. On exposed rocks there are oysters, limpets, barnacles and sea snails. In the crevices there are sand crabs and sea urchins.

The importance of shellfish to the first Australians is underlined by the virtually continuous chain of shell middens round the coastline of the continent. Some are quite small—the product, perhaps, of a single meal or an overnight camp. Others are the result of hundreds or even thousands of years of feasting. In parts of Tasmania these layers of whitened shells extend for several kilometres along the beach. On the western side of Cape York in Queensland there are shell middens that are ten metres high—shell mountains made up of a single species of cockle.

The Cape York shell middens have long intrigued archaeologists, for a number of reasons. The construction of such huge piles must have involved many people, in the later stages, in considerable labour, just in carrying shells to the top. But the purpose of all this effort remains a mystery. The middens do not seem to have been used as burial mounds, for no human remains have ever been found in them. They might have offered refuge from floods, but there is equally suitable high ground nearby. The present people in the Weipa area, who are custodians of the middens, cannot provide any explanation.

Another odd thing about the mounds (although this does have a logical

The shell middens which
extend around the entire
coastline of Australia are a sign
of the importance of this food
resource to hunter gatherers,
and of the industry of the
women who collected them.
*This midden of shells, on
Fraser Island in Queensland, is
the result of hundreds or even
thousands of years of feasting.*

explanation) is the fact that they are located several kilometres from the sea, in woodland. This does not, however, mean that the people who made them carried the shellfish all that way inland before eating them. Carbon dating of materials from the base of the middens shows that they were begun more than a thousand years ago, and at that time they were in fact close to the beach. In the intervening period siltation from nearby rivers has raised the level of the land, and the sea has retreated.

The middens around the Australian coastline are interesting in themselves, as testaments to the enormous industry of the women who collected the shells, but they also provide a clue to the manner in which the continent was probably settled. The first arrivals were, by definition, coastal people, for they came from islands to the north and west, and landed on beaches across northern Australia. The abundant food resources of the coast would at first have supported the new populations, even as numbers grew. Even when the need arose for groups to move away and find new living space it is likely that they would have expanded first along the coast, which provided a regular continuation of a familiar environment. Movement along beaches and coastal plains can be comparatively rapid, and although Tasmania is a very long way from northern Australia it is quite possible that by spreading round the coast the early colonists may have reached that distant territory within a surprisingly short time— perhaps one or two thousand years.

Sooner or later, of course, people began to move inland, into new environments. In Arnhem Land we can find the descendants of some of those groups in the large community based on the township of Maningrida, on the banks of the Liverpool River. Although they spend much of their time in the settlement, an increasing number of family groups return to their tribal lands for long periods, up to several months in each dry season, in what has come to be called the 'outstation movement'. At these times they live a fully traditional existence, hunting and gathering wild food, teaching the children bush lore and tribal myths, making baskets and other useful artefacts, creating bark paintings and carvings, and holding the rituals and ceremonies which have been the backbone of their culture for tens of thousands of years. At these times they, too, provide a living window into the lives of the early Australians.

On the flood plains of Arnhem Land, below the great sandstone escarpment that forms a massive red backdrop, all life has a pronounced seasonality, because of the monsoon. In the wet season, in summer, huge banks of clouds roll in from the northwest, and rains pour down every afternoon. Rivers and billabongs overflow and turn the plains into a gigantic swamp. Grasses shoot up two metres or more. Hunting is virtually impossible, and people stay in their shelters and live on plant foods.

As the monsoon recedes, and the land begins to dry out, a time of plenty opens up. Magpie geese in tens of thousands descend on the lagoons to nest and breed. They provide targets for the spears and nets of the men, edging their way through the reeds in their dugout canoes. Others weave long, cylindrical baskets from flexible twigs and lower them into the rivers to trap the fat barramundi, a fish that grows to ten or fifteen kilograms.

But the really productive activities at this time are all to do with gathering, not hunting, and here again it is the women, with their hands, who provide most of the subsistence. They wade out to the floating nests of the magpie geese, up to their necks in water, and collect the eggs in thousands. They dig up the roots and bulbs and stems of waterlilies in the lagoons, and catch small fish in the shallows. In some pools there are colonies of large, fat, nonvenomous file snakes (so called because of their rough skin). The women locate them with their feet, then throw them out on to the bank.

Along the sandy ridges and in the woodlands the women catch a whole range of small animals. They dig eight or nine different kinds of lizards out of their holes or from under logs, killing them with a practised swing against a tree. There are pythons curled up in sunny patches, and small possums and gliders sleeping in tree hollows. Occasionally an echidna, or spiny anteater, is found pushing its way through the grass. This is a delicacy, because of the layer of fat beneath its coat of spines. Tiny black shapes buzzing around a dead trunk reveal the location of a nest of native bees, oozing honey.

On the trees there are various berries, seeds, pods and fruits, including figs and wild 'plums' and 'grapes'. In the sandy ground there are succulent yams of different kinds, easily extracted with a digging stick. (The women sometimes cut the yam off the bottom of the vine, leaving a small fleshy portion attached, and push the stem back into the hole. This is perhaps as close as the Australians ever came to growing food—a subject that is explored in Chapter 7: Changing the Menu.)

As the dry season takes over, and the lagoons and billabongs contract, other foods become available, including tortoises, freshwater crocodiles and eels. Huge numbers of waterbirds become concentrated on the few remaining patches of water, to feed on worms and the bulbs of spike-rushes, exposed on the drying mud flats.

Even later, when the tall grass has dried out, the plains are fired. The purpose is threefold: it drives out wallabies, emus and other game that can be speared by the encircling hunters; the rapid regrowth of fresh green shoots attracts game on to the burned patches, where they can be more easily speared; and, finally, the clearing of the bush simply makes it easier to walk through. The plains of Arnhem Land are dotted with fires each dry season, and recent studies of aerial photographs show that over a cycle of three to five years every bit of land is deliberately burned at least once. This burning—virtually a house-cleaning—is carried out in many parts of the continent, and has been an important element in the adaptation of hunter gatherers to the Australian environment. It is also part of a much wider question of environmental modification throughout the Pacific basin, and is discussed in our final chapter.

Another environment that was important in the early colonisation of Australia was the tropical rainforest. This complex, crowded plant community has been associated with the peopling of the Pacific from the very

The lagoons and flood plains of Arnhem Land are a rich environment for hunting and gathering. *Towards the end of the dry season, lagoons like this one become a focus of bird and animal life.*

During the wet season in Arnhem Land the people make wicker fish traps to catch barramundi in the rivers. *This one is ready for action at one of the outstations of the Maningrida community, on the Liverpool River.*

A bark painting illustrates the way the barramundi enter the wicker fish traps. *This painting is by an artist of the Maningrida community in Arnhem Land.*

beginning of that story, when the first humans, *Homo erectus*, came into the area and settled into the jungles of Java.

Even during the recent ice ages, when conditions were generally cooler and drier, tropical rainforest still covered much of lowland New Guinea, and extended down the east coast of Australia. At times of low sea level, when New Guinea was joined to northern Australia, there would have been a continuous belt of rainforest down the northeast coast of Greater Australia. The key indicator of this is the continued existence of the same species of animals in both of the now widely separated areas of rainforest: tree kangaroos, amethystine pythons, cuscus (large possums) and cassowaries (large emu-like birds).

When the first people from Southeast Asia crossed the last water gaps and reached Greater Australia, some probably landed on the northwest coast near Arnhem Land, as we have already suggested. But others undoubtedly headed eastwards, into the rainforest along the Pacific coast of New Guinea. Once adapted to this specialised environment some moved south, into the rainforest along what is now the coast of Queensland. When the seas rose at the end of the ice age and cut off Australia from New Guinea, the rainforest dwellers were isolated in their southern tropical habitat. Here they lived on, a distinctive group of hunter gatherers, in some ways quite unlike other Australian people.

One of the fundamental differences was in physical stature. Like rainforest dwellers elsewhere, the Australians had to adapt to the dense, crowded environment of closely set trees, vines and dangling lianas. Tall people would find this tangle difficult to move through, and would be disadvantaged in the hunt for small, elusive animals. The result was that in succeeding generations smaller people were 'selected' (in the sense of Darwin's theory of natural selection). Adult rainforest men were, on average, less than 158 centimetres (5 ft 2 ins), and women were less than 153 centimetres (5 ft).

Because of the closed environment in which they lived, the rainforest people had to devise new ways of obtaining their food. They were not short of edible plants, but some species of Southeast Asian origin were, as mentioned earlier, quite toxic. So they became expert at removing harmful alkaloids by leaching them in the rushing forest streams. They also found an efficient way of using the macadamia nuts of the rainforest (the only native Australian plant food that has ever achieved international acceptance). This nut has a smooth, round and remarkably hard shell, and shoots away like a bullet from most attempts to crack it. But in a few corners of the rainforest there can still be found small rocks with rows of nut-sized indentations in them, where people lodged the nuts and cracked them with a hammer stone. The depth of the holes is an indication of extremely prolonged usage of these useful artefacts.

Since the rainforest floor was dark and covered with leaf litter, animals could not be tracked, and because of the tangle of vegetation they were hard to spear or hit with throwing sticks. So the rainforest people developed, as did others in the same environment elsewhere in the world, many ingenious traps and snares for catching game. Most of these—like the culture of the rainforest people themselves—have now disappeared, decomposed by the humid atmosphere of the rainforest. But on the coast nearby there are more permanent examples of the trapping techniques of these people.

These are stone fishtraps, built from rocks and boulders in the form of squares or circles. The basic principle of what one anthropologist has described as an 'automatic sea-food retrieval system' was simple, efficient and, after the initial work of construction, labour-free. An enclosure with walls up to a metre high was built on the tidal flats, next to the mangroves, where fish were known to feed at high tide. The tide covered the enclosure for some hours, and when it fell fish and other marine animals were trapped inside the stone walls. All the people had to do was to come along twice a day and collect their catch: school fish like mullet and tailor, barramundi, stingrays, squids, mud crabs and crayfish. Even the oysters

Fish traps, built from boulders on many parts of the Australian coastline, are filled by the sea at high tide, and retain their catch when the tide goes out. They are an efficient and labour-free method of obtaining food on a daily basis. *This trap, on Hinchinbrook Island in Queensland, encloses an area of several thousand square metres.*

The people who lived around Sydney Harbour had access to a wide range of foods, especially fish and shellfish. *This is one of the illustrations made by J. H. Clark in Sydney and published in 1813.*

which grew on the rocks were a kind of edible cement. There were rocks that could be removed to make exits, so that surplus fish could be allowed to survive for another day.

Fishtraps which worked on these principles were used all round the Australian coastline, and on many inland rivers, but none were as extensive as those built by the rainforest people of Queensland. There is one such trap still standing on Hinchinbrook Island, south of Cairns, which covers an area of several thousand square metres. It has a number of enclosures, built at different levels, so that at normal tides only certain parts are exposed, to act as traps. Other sections are in deeper water, and only exposed at very low tides. At these times of maximum catch, people from some distance around gathered for ceremonies, so relating their social cycles to those of nature. Besides the conceptual skills which these fishtraps represent, it seems delightfully ironic that the smallest Australians should have made the largest artefacts on the continent.

Following the route of the early Australians as they moved around and down the Pacific coastline, we come to one area which was particularly rich in food resources, as well as habitation sites in the form of caves and overhangs. The first arrivals on this scene would have found high sandstone cliffs looking out across a plain to the sea, with rivers cutting deep channels back into the hinterland. The dry eucalypt forest along the clifftops contained large grey kangaroos, swamp wallabies, possums and koalas, goannas and pythons, emus, scrub turkeys, parrots and fruit bats— all good hunting. The rivers and estuaries had fish and crabs, oysters and other shellfish. The climate, even during the ice ages, must have been mild and not too dry.

But as many people as may have paused here on their way south, and perhaps made this area their home, their numbers must have been small compared to the population which developed after the ice age ended, and the sea rose to its present level. The ocean now lapped at the foot of the great sandstone cliffs. The river valleys had become deep inlets, with many sheltered arms and mangrove-lined estuaries where fish could breed and multiply. The climate had warmed up, vegetation was spreading, and life of all kinds flourished, both on the land and in the sea. Whales and seals, huge rays and sharks swam off the coast.

We know this because the people who lived in the area created an impressive body of rock art—outlines of whales, sharks, kangaroos and emus, pecked into smooth rock surfaces near the ocean. The middens around all the inlets and on the headlands also mark their presence here, and display the variety of their diet, obtained by fishing from dugout canoes and hunting along the clifftops. They contain fish of all kinds, abalone, oysters and mussels, fruits and bulbs, and the bones of many animals and reptiles. The people here clearly found this great system of cliffs, sheltered inlets and quiet backwaters an attractive place to live. It is interesting to reflect that their assessment was echoed by those who came after them, because this was the place that Governor Arthur Phillip chose as the site for the first European settlement in Australia—Port Jackson or, as it was later named, Sydney.

While coastal environments such as that just described obviously provided a good living, and were well populated by the first Australians, the continent in fact did offer some other habitats which were even more rewarding for hunter gatherers. These were the inland river systems, with their chains of billabongs and swamps. These, in the end, supported perhaps the densest populations of people that ever existed in prehistoric Australia. The most important of these was the Murray-Darling system, which together with the Murrumbidgee River drains the entire southeast third of the continent, west of the Great Dividing Range.

These rivers wind across the flat plains, and when overflowing with spring run-off from the mountains they break their banks and fill thousands of billabongs and small lakes. In the warm, nutrient-rich nurseries life proliferates with astonishing fecundity. On the waters there are vast flocks of ducks of a dozen different species, swans, ibises, cormorants, geese, egrets. In the pools and rivers there are monster Murray cod, weighing up to a hundred kilograms, golden perch, tortoises, crayfish, mussels, frogs. On the banks there are grey kangaroos, emus, small wallabies, goannas, snakes, and in the trees possums, gliders, eagles, hawks and owls. The silt-laden flats beside the rivers are crowded with small, juicy yams and bulrush bulbs.

The hunter gatherers who came to this cornucopia found life so easy that they virtually became sedentary. They lived along the banks, under the great river red gums, and obtained their fill with very little effort. There are accounts of men returning from the duck-breeding rookeries with so many eggs that their canoes overturned. One of the first European observers of this scene saw a group of people breaking camp, and asked them why they were leaving, when there was so much food to be had. They said they were going south to find something different to eat—they were bored with the food! And we know that this bounty had persisted for at least 30 000 years, because the archaeological finds at Lake Mungo (described in the previous chapter) reflect a similar situation around the Willandra Lakes.

One result of this availability of food was abundant leisure time. This, in turn, provided the opportunity for ceremonies, dancing, and expressions of all forms of artistic creativity. Although, as already explained, most artefacts made by Australians perish very quickly, the observations and collections made by the first Europeans to travel the Murray-Darling system show an astonishing variety of shields, baskets, jewellery and personal decorations. Some of this individuality—even the styles adopted by particular groups in the area—can only be described as high fashion.

Beyond the Murray-Darling system, towards the heart of the continent, the situation was quite different. According to archaeological evidence, there were people in the centre at least 20 000 years ago. At this time the last ice age was at its maximum. The arid third of the continent was even drier, and considerably colder, than it is now. And yet people occupied this region, too, and adapted to its harsh conditions—so successfully that their

Birds which flock to breed on the Menindee Lakes include cormorants, pelicans, ibises, ducks and swans. *Young cormorants dive for safety at the approach of a boat.*

way of life survived, virtually unchanged, right down to the present. The image of the desert tribesman in fact became the stereotype of the aboriginal Australians.

By contrast with the inhabitants of the Murray-Darling basin, the desert people were few and widely scattered. This is not unexpected, since there is a good relationship between population density and rainfall, or available surface water, and over much of central Australia the average annual rainfall is less than 200 millimetres. There are no permanent rivers, and even when lakes are filled after rains the water is too salty to drink. So tribal areas were huge, with family groups spread out far from one another for most of their lives. It is calculated that the desert people, occupying some thirty per cent of the continent, never made up more than about ten per cent of the total population. In many ways, however, they lived like other Australians—although sometimes, in periods of drought that might last for years, on a knife-edge of survival.

When they go out from their camps into the desert the men hunt kangaroos and emus with their spears—but the women find most of the food: lizards, snakes, honey-ants, witchety grubs and small marsupials. Plant foods are limited in the desert, apart from yams and a few small

During the wet season the inland rivers of Australia flood large areas of the plains, creating lakes, lagoons and billabongs that attract huge concentrations of bird life. *This is one of the Menindee Lakes, in western New South Wales, fed by the overflow from the Darling River.*

fruits. The seeds of various grasses are an important staple; the women grind them into flour and make a kind of pancake. When people began pushing into the desert from the coastal regions, the process of learning how to use these seeds was one of the ways in which 'botany began anew' in Australia.

Most outsiders see the desert as alien and hostile, a dangerous place for those who cannot adapt to its conditions. The desert people, who have adapted to it and know its moods, see it as a garden, in which they move with ease and security. But it is the outsider's fear of the desert that has protected the desert people from invasion and destruction, just as it has protected hunter gatherers in other parts of the world. And that has permitted the survival, virtually intact, of the culture, the ceremonial life and the artistic traditions of the Australian desert people.

Examples of this, little known even to aboriginal Australians in other areas, are the little sculptures called 'toas' which we showed for the first time in our film series. These come from one featureless area near Lake Eyre, and they are unusual art objects in that they combine the mythical or supernatural world of the Diyari people who made them with an extremely practical system of signposting.

There are few trees or other available materials in the area, so the Diyari took short sticks and modelled clay shapes on to them, and then painted the finished sculptures with pigments made from red, yellow and white ochres, and black charcoal. In their constructions they sometimes included natural objects such as feathers, shells, a stone, or a lizard's foot. They made hundreds of them, in innumerable shapes, but none taller than about twenty centimetres.

Nearly all the toas are symbols of places around Lake Eyre—a dry watercourse, a clump of trees, a rock-hole where water can be found—or of events, supernatural or real, which occurred in the past. And their practical use was as signposts. When people were going on a journey they made a toa of the place they were heading for, and left it in their camp to tell others where they had gone. Toas were signposts to the real world, and to the realm of the spirits.

Very little is known about the origin of the toa tradition, or how long people went on making these little sculptures. They are known only from a single collection, acquired at the end of the last century and now in the South Australian Museum. In terms of Australian art, they may represent a quite new form of expression by desert people. What is also interesting is that not far from where the toas come from there exists another form of expression altogether, which may well be the very oldest example of human art in Australia.

These works, which are among the most mysterious and unusual forms of art discovered anywhere, were made by desert people who went down into a huge complex of limestone caves beneath the Nullarbor Plain to obtain lumps of flint, which are constantly eroding out of the walls. But they also went deep into the darkest recesses and made complex designs on the walls by drawing their fingers across the soft limestone. In some

Toas are small, very unusual sculptures of wood and painted clay, made by the people who once lived around Lake Eyre. They were used as markers, and placed in the ground by travellers to indicate where they had gone. Some toas were abstract *(a)*, others quite representational. Some included natural materials *(b)*, such as a lizard's foot or a mussel shell. The dots and lines on toas *(c, d)*, could stand for natural features in the desert, such as trees or watercourses. *These are from the collection in the South Australian Museum in Adelaide.*

places they incised lines into harder surfaces. Carbon dating of charcoal found near the markings — perhaps from crude torches — suggests an age for this art of at least 20 000 years. Similar markings in European caves, dated to about 30 000 years ago, have been described as perhaps man's earliest artistic endeavours. The Koonalda markings may therefore be the oldest art in Australia.

In any account of the first Australians and their conquest of a continent — 'the triumph of the nomads' — there is one population which occupies a special place. The story of the people of Tasmania is a saga in itself, a tragic chapter in human history; here we can only outline it briefly.

No one knows exactly when the first people crossed the exposed land bridge between Australia and Tasmania, but it was at least 23 000 years ago, when the ice age sea level was extremely low. We can say with certainty, however, when the last people made that crossing. The rising seas at the end of the ice age cut the land bridge about 9000 years ago, and

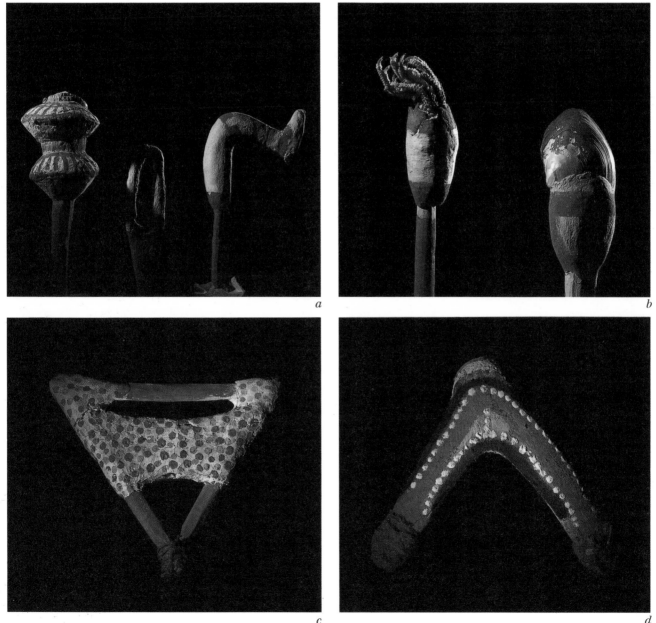

a

b

c

d

The marks made by human hands on the soft limestone walls of Koonalda Cave in South Australia are estimated to be about 20 000 years old, and may be the oldest art in Australia. *The marks are thought to have been made by people going down into the cave to obtain large lumps of flint, which are continually eroding out of the walls and roof.*

TERRE DE DIÉMEN.

There are still many descendants of the Tasmanian people, but with the loss of their way of life and culture, the world lost a unique opportunity to learn more about the human condition, under very particular circumstances. *This striking portrait was made by Nicholas Petit.*

the stormy expanse of Bass Strait prevented any further contact. Thus was created the most isolated group of people the world has ever known, cut off from the rest of humanity for nearly 10 000 years, until Abel Tasman's visit in 1642.

The archaeological evidence, and the state of Tasmanian society when that long isolation was breached, tells us that this southerly way of life was an uncomplicated one, based on what was perhaps the simplest set of material possessions humans have ever relied upon. The people lived in low bark shelters, built over hollows in the ground. Their tools consisted of wooden spears (without stone points, and without the woomera, or spear-thrower), a throwing stick and a few stone choppers and scrapers. Their only containers were baskets made from woven rushes or seaweed. The men used a canoe-shaped raft made of rolled-up sheets of bark, and women who dived for abalone wore a small basket and carried a short, stout stick to prise the molluscs off the rocks. Shell necklaces were their only personal adornments. And that was all—a total of less than a dozen kinds of articles.

The Tasmanians lived a generally coastal existence, as the extensive shell middens show. These also contain the bones of seals, and over-hunting may have contributed to the eventual disappearance of elephant seals from Tasmania. The finds in Kutakina Cave, on the Franklin River, show that some people were living there, well inland, 20 000 years ago. They moved up the river valleys in summer to hunt wallabies and other animals, and returned to the coast in winter, when ice and snow covered the high plateaus.

When the Tasmanians were first discovered, their lack of material development, compared to what had occurred in mainland Australia, suggested to some observers that the long isolation had caused a slow cultural decline. It may well be, however, that they were simply representative of the kind of people widespread on the mainland at the time of their separation. The development of more sophisticated tools on the mainland, for example, was probably a response to the much greater environmental changes that took place there after the end of the ice age. Some new types of tools may even have been imported from outside Australia, just as the dingo was.

Today, unfortunately, the Tasmanian way of life has gone, although the people themselves have survived in the many descendants of the last full-blood men and women, who had all died by 1876. The myth of their extinction, and the notion that full-bloodedness is somehow essential to ethnic identity, still fuels misunderstandings and bitterness. The conflict between pride and guilt tends to make us forget that in the destruction of that fragile, isolated culture we have lost an important link to the human past, an irreplaceable window into the world of the hunter gatherer. It was an opportunity to learn something about ourselves which can now never be repeated.

However, in the general reappraisal in recent years of the achievements of the early colonists of the Pacific basin in general, and of Australia in particular, one myth above all has been exploded. That is the image of the hunter gatherer, the 'noble savage', living in perfect harmony with nature, in a timeless, unchanging continuum. We know now that in fact the first people in this part of the world had a dynamic, ever-changing relationship with their environment, in which they both adapted to and actively modified the world around them. In a sense, they made the world they lived in. The first Australians were no exception, and may yet be shown to have had a greater impact on their continent than any of the other peoples of the Pacific. As stated earlier, this intriguing concept is to be explored more fully in our final chapter.

In the meantime, it is well to remember that the migrations which peopled the Great Southern Continent during the ice ages were only part of that general movement out of Asia which was to have such far-reaching consequences. And even as the first explorers were pushing south into the deserts of Australia, others were heading north, into another unknown and even more inhospitable region—Siberia.

The tool kit used by the Tasmanians contained the fewest number of items ever used by any human society. *This illustration, by Nicholas Petit, shows two throwing sticks, spears, baskets and a shell necklace.*

TERRE DE DIÉMEN.

ARMES ET ORNEMENS.

1. Sagaie. 2. Casse-Têtes. 3. Vase à Eau. 4. Sac de Jonc. 5. Collier de Coquillages.

The valley of the Franklin River in Tasmania was one of the areas occupied by the Tasmanians at the coldest period of the last ice age. *This shows the rocky limestone walls near Kutakina Cave, which was occupied 20 000 years ago.*

The Tasmanian people, cut off by the rising sea level at the end of the last ice age, 10 000 years ago, survived the longest isolation in human history. *This illustration was made by Nicholas Petit, the French artist with Baudin's expedition of 1802.*

4.

Into the Deep Freeze

O ur knowledge of the first human migrations north from Asia into Siberia during the ice ages is still as hazy as the horizon of the vast expanse of tundra itself, vanishing over the curve of the earth. There are few archaeological sites in northern China, Korea or Siberia, and they contain little more than some stone tools and the bones of hunted animals. Human remains, that could give us clues to the development of those early explorers, are virtually non-existent. Fortunately, there are still a number of hunter gatherer groups living in areas of northern China and Siberia, and their life-style provides a vivid guide to the adaptations, both physical and cultural, that had to be made by the first people to venture north, into the deep-freeze zone of the planet.

The environments the north seekers entered, and the climatic conditions they encountered, were in utter contrast to those experienced by the groups moving south, into the tropical rainforests of Southeast Asia and the arid plains of Australia. After leaving the temperate woods of northern China they would have found themselves traversing the gloomy fastnesses of the taiga—the mighty tract of tall, dark, closely spaced trees that extends in a single block across Europe and Asia, from Scandinavia to the Pacific Ocean. The taiga consists almost entirely of evergreen conifers— fir, spruce and larch—interspersed with stands of deciduous birch.

A few mighty rivers cut through the taiga on their way from the interior of Siberia to the Arctic Ocean. One, the Lena, is among the world's longest, and when its frozen surface melts in spring it carries a greater volume of water than the Amazon. Another is the Amur, which flows east and south to empty into the Pacific. These rivers are flanked by the luga—a low-lying grassy prairie produced by regular flooding. The luga is of ecological importance because it is the only natural grassland in an otherwise unbroken belt of forest. It also provided convenient passageways north for human travellers.

At its northern edge the taiga thins out into a woodland of stunted firs,

Northeastern Siberia, with its forests of conifers, mighty rivers and treeless tundra, is one of the last great areas of untouched wilderness left around the Pacific basin. *This is a gorge leading to the Lena River, which drains into the Arctic Ocean.*

71

larches and scattered willows. This in turn gives way to the seemingly endless expanse of the tundra, which stretches to the shores of the Arctic Ocean. The tundra is flat and featureless, covered with a low, treeless carpet of dwarf shrubs and hardy plants which spend much of their existence buried beneath ice and snow. In winter this is a desolate wasteland, chilled by the icy winds which swirl around the North Pole. Beyond the tundra lies the shore of the Arctic Ocean—the engine of this gigantic refrigeration system.

The whole enormous region of eastern Siberia has barely been touched by civilisation, and remains the last great bastion of natural wilderness on the margins of the Pacific Ocean. During the ice ages it would have been colder and drier, but otherwise not very different from the way it is today. From its few scattered, tenacious groups of traditional inhabitants we can get a picture of what life must have been like for the first humans who lived in one of the world's harshest environments. The records of their way of life come from explorers and anthropologists who went into the area in the nineteenth century, before the nomadic people there had been in contact with any kind of modern technology. The most detailed accounts are of the Oroqen people of northern China.

The Oroqen once ranged across a wide area, including parts of Siberia, but are now confined to the Chinese province of Heilongjiang and the eastern part of Inner Mongolia, around latitude fifty degrees north. Their territory is wild and rugged, cut by many rivers—one of which, the Amur, forms the border with Siberia. The Oroqen live in an area of forests, some almost pure stands of pine, others containing larch and birch. There are also mixed forests of oak, maple and poplar, and a few patches of grassland to break up the otherwise continuous ranks of trees.

The whole area is rich with food resources: bears, tigers, boar, moose, roe and red deer, lynxes, martens, minks, rabbits, otters and squirrels. In summer there are swans, cranes and other migratory birds. The rivers and lakes carry salmon, sturgeon and carp. The forest understorey yields hazel nuts, small fruits and berries, fungi and edible roots. The seasons are extreme, with hot and humid summers and frigid winters, in which the temperature falls as low as minus forty degrees Celsius, and the whole countryside is buried deep in snow.

This was the environment encountered by the first humans to begin the movement north, away from the temperate woodlands and plains of China. To survive they had to adapt to its conditions. Inevitably, they became expert hunters with spears, bows and arrows, harpoons and traps. The major element in their diet—unlike that of the people who moved south, into Southeast Asia and Greater Australia—was meat. The warm, fur-lined clothing essential for survival in this climate also came from hunted animals, and to obtain the number and variety they needed these forest people were obliged to be constantly on the move, hunting and trapping in all seasons. Small groups, often single families, ranged widely to take advantage of local availability of food.

Men and women had separate roles in the constant search for food.

The physical adaptations made by the people who first went to live in Siberia's extremes of climate produced what we know as the Asian face. *These and other historic photographs in this chapter were taken in the late nineteenth century by expeditions from the American Museum of Natural History.*

Hunting was normally carried out by the men alone, but acting co-operatively. Some would circle behind the game and drive the animals from cover towards the bowmen and spearmen. In the mating season of the deer one would use a wooden deer whistle, which sounded like a stag. Other stags, assuming a challenge to their harems, would charge towards the source of the sound and into the range of the spears and arrows. The women and children foraged separately for fruits, berries and nuts.

Fishing was one of the few food-gathering activities that men and women carried out together. When fish were running, young men cut poles and drove them into a stream bed at a suitable point, making a line of pilings from one bank to the other. The women wove willow twigs into mats and long, tubular baskets, open at one end. The mats were placed against the pilings to form a barrier with a number of openings, into which the baskets were fixed. As the baskets became heavy with fish the men went out and hauled them ashore. In another technique, practised in lakes

Among the most important sources of food for the early Siberians were the salmon and other fish in the rivers, which they caught with traps made of twigs. *These were made by the Yukaghir people on the Korkodon River.*

Whitefish were caught in large numbers during their spawning runs from the ocean into the Nalemna River. *The women often worked for hours with bare hands in subzero conditions, cleaning fish for drying and freezing.*

The typical summer shelter right across Siberia was the conical tent, covered with birch bark or skins. *Such tents were ideal for nomadic hunter gatherers, because they were expendable and easily replaced.*

on calm summer nights, they used flaming birch bark torches to attract fish into the shallows, where they could be speared. Much of the game and fish was eaten immediately, but some, especially in the increasing chill of autumn, was dried and stored in leather bags for the winter.

The forest people had access to a wide range of natural materials for the manufacture of hunting weapons, tools and articles for daily use. Along the rivers they found several kinds of hard stone, suitable for making knives and scrapers. Where particular types of stone were not available locally they were traded in, sometimes from considerable distances. Because of the great variety of animals in the region there was always a good selection of skins, horns and bones for different purposes. Among the unusual pieces of hunting equipment they used were eye protectors, made of plaited hair. These acted like sunglasses in reducing the reflected glare from the snow.

Clothes were made almost exclusively from animal skins. The men saved the hides of all animals killed while hunting and brought them home to the women, who did the work of tanning. Some skins were treated with the fur still on them, while others had the fur stripped off. The women first dried the skins, then scraped all traces of flesh away. Then they reduced them by prolonged kneading to soft, supple leather. Finally the skins were smoked over a fire; this process prevented them from stiffening when they got wet.

Clothes were generally worn with the fur on the inside, to provide the maximum insulation. They were loose-fitting, to allow warm air trapped against the body to circulate. This helped to prevent excessive perspiration from dampening the clothes, and so destroying their insulating properties. The upper garment, a kind of parka, usually had an attached hood,

People have been living in Siberia as hunter gatherers for tens of thousands of years, and many, like the Yukaghir, still follow that ancient tradition. *The hunting bow and the skin-covered tent were both taken into North America by the first people to settle that vacant continent.*

although some individuals preferred a separate fur hat. Boots were made of very tough skins, and were sometimes further lined with grass to make an additional layer of insulation. Mittens were carefully made, because of the danger of frostbite. Clothes were normally sewn with coloured strips of skin or gut, and decorated with vegetable dyes. There was one other item of fur which everyone had: a sleeping bag. This was used outside in the summer, and inside in winter.

Each family lived in a tall, pointed shelter, based on a framework of thin poles arranged in a circle and lashed together near the top. In summer the frame was covered with sheets of birch bark, which gave quite adequate shelter from sun and rain. In winter the frame was covered with a tight-fitting leather tent, made from sewn skins. The stretched surface shed snow, and provided good insulation against loss of heat from the small fire and crowded human bodies inside. Such a shelter placed no restraints on the movements of these nomadic hunters. In summer the family simply abandoned it when they moved to a new hunting area, and in half a day built a new one. Even if they were obliged to move in winter all they had to carry was the skin cover.

The mixed forests provided the people with an inexhaustible supply of timber and bark for a hundred different uses. Their range of wooden artefacts included aids to snow travel that closely resembled skis: strips of split birch timber about two metres long and twenty centimetres wide, pointed at both ends, with the front ends turned up. Some even had a layer of deer skin on the running surface to make them move more freely over the snow.

These people discovered many uses for birch bark. In spring, when the smooth white outer layer was thin and flexible but very tough, they cut it off the tree trunks in sheets. From this they made shields, drums, boxes, buckets, bowls, wallets and quivers for their arrows. They decorated the finished objects by stitching, painting and incising the surface in such a way that layers of different colours were revealed. With some articles, such as boxes, they added black and white porcupine quills, sewn on in complex patterns.

One of the most important uses of birch bark was for making canoes. The bark was applied in sheets over a flexible pine frame and sewn together with willow twigs. This produced a light and highly manoeuvrable craft up to about four metres long, which was easily paddled or carried over rapids. (There was a striking resemblance between these canoes, shelters and quill-covered boxes and the birch bark canoes, skin-covered teepees and quill boxes used later by the North Americans. As we shall see in Chapter 5, this was more than just the coincidental use of similar materials by people in similar environments.)

The social structure of these bands of hunter gatherers was quite loose. Four or five family groups—perhaps related through a single male ancestor—occasionally joined into one economic unit. Such a unit would contact others to trade in scarce commodities, arrange marriages, and organise ceremonies. These were conducted by shamans, and sometimes, in order to celebrate the productivity of the land, involved the ritual burial of animals killed in the hunt.

As people expanded even further north, life during the ice ages became increasingly difficult. Not only were climatic conditions more severe, but the range of foods and raw materials was much narrower. The human adaptations demanded in these new environments were to influence northern and northeast Asian people both physically and culturally. One of the most profound sets of modifications produced what we now recognise as the Asian face.

The physical differences between major divisions of humanity—Africans, Europeans, Asians, Melanesians—have come about through a series of both short and long term biological adaptations to different environments. Some features, like teeth, have been fairly stable in their general shape and proportions for about a million years. Other features, such as brow ridges and chins, have changed noticeably over the same period.

We can trace some of these underlying bony features a long way back in time, but those characteristics that we now recognise as African or Asian or European—outward shape, skin and eye colour, hair type—do not leave a fossil record. Bodies have been found preserved in peat bogs in Denmark, frozen in ice caves in the Andes, embalmed in Egyptian tombs, but none of these examples of past humans is more than a few thousand years old. Consequently, they all look quite modern. Questions as to why Europeans have white skins and Africans black, or why aboriginal Australians are generally tall and have long thin legs, cannot be answered or even tested with real evidence.

Some of the differences we see appear to make good adaptive sense. The

The bark of the birch tree was one of the most useful materials available to Siberians. *It was thin but tough and flexible, and easily cut off in sheets.*

Birch bark was used for making containers and storage boxes of all kinds. *These were made for sale at the annual fair at a village on the Lena River.*

fact that white Australians have the world's highest incidence of skin cancer, while black Australians rarely suffer from the same problem, suggests that black skin is an advantage in hot, very sunny countries, where solar radiation in the form of ultraviolet is intense. In such climates, pale skin derived from a European background is a disadvantage. In northern Europe, on the other hand, there seems to be good reason for thinking that light skin enables more ultraviolet solar radiation to be absorbed, to synthesise vitamin D in the body. (Lack of vitamin D leads to rickets.)

But while it is possible to relate single characteristics, such as skin colour, to specific adaptive conditions, it is difficult to show that all the distinguishing facial features of Europeans or of the original Australians are the result of a single set of environmental conditions. The features of Asian people are, however, rather easier to interpret in this way, because so many of their special characteristics appear to be the result of a single

Birch bark tents and canoes beside a river in Siberia illustrate the value of this natural and easily obtained material to the people of the north. *Birch bark canoes were light enough to be carried easily by one man.*

adaptation, to cold.

Overall, the Asians have proved to be an extremely adaptive group, and their success has led to a huge expansion of their original range. The biological term for them—mongoloid—refers to a physical type rather than someone from a geographical area in Asia. The term is also applied to people of this unmistakably Asian type, even though they now live in Indonesia or even South America.

Mongoloid people display a set of cold adaptations that are similar to those found in animals from polar or high altitude habitats. They tend to be short and compact, with a large body and relatively short limbs. This means that they have a smaller surface area in relation to their volume, which reduces heat loss. People who live in hot deserts, on the other hand, tend to be tall and thin-limbed, which increases their surface area and helps to dissipate heat. The shape of the head is important in this context. The smallest space into which any given volume will fit is a sphere, and in their braincase the mongoloids come closer than any other humans to

The flattening of the face is one of the most obvious adaptations to cold among northern people. *The shortening of the nose reduces the risk of the tip freezing.*

having a perfectly globular head.

Mongoloid faces are small, broad and flattened. The short nose (like other short extremities, such as the fingers and toes) minimises the risk of freezing. The flattened nose with narrow nostrils slows down the intake of cold air, and helps to warm it before it reaches the lungs.

Another distinctive mongoloid adaptation is in the amount and distribution of body fat, which serves as both insulation and reservoir of energy. Mongoloid fat is distributed generally over the body surface, so that all parts are protected, but two areas of the face are particularly favoured. The cheek bones, because of the flattened facial structure, are sharply angled outwards, and a special pad of fat overlies and protects them. The other area of facial fat is perhaps the most characteristic facial feature of mongoloids—the epicanthic fold, or so-called extra eyelid. This is a flap of skin, filled with fat, which covers the upper eyelid, so that the upper eyelashes appear to emerge from it. This fold helps to insulate the upper part of the eyeball, and by narrowing the eye slit it also helps to reduce snow glare. The yellowish skin colour of Asian mongoloids is partly a function of the layer of fat just below the skin, but the generally light skin tone seems to have a function similar to that of other light-skinned northern people—that is, to promote the absorption of ultraviolet radiation and the synthesis of vitamin D.

The hair of mongoloid people is thick and lies flat on the scalp, forming good insulation. This is because the individual hairs are round and straight. Africans have hairs that are flat and tend to form spirals; this produces a loose, curly head of hair which allows air to circulate and thus remove heat. The need to maintain protective insulation may explain why mongoloid people have a low incidence of baldness. Their hair is also invariably black or very dark. This promotes the absorption of solar radiation and helps to warm the scalp in cold weather. It is not so clear why mongoloids have such sparse body hair, although as far as the face is concerned it has been suggested that hairlessness reduces the risk of

Yakut people at a village fair in central Siberia show the range of Asian faces. *Everyone puts on their best clothes for the one social gathering of the year.*

The fatty cheek pads are another cold-climate facial characteristic. *Besides a general layer of fat under the skin, the Yakut have large bodies and short limbs, to reduce heat loss.*

The fat-filled 'third eyelid', or epicanthic eye-fold, is typical of northern people. *This feature protects the upper part of the eyeball, and cuts down snow glare.*

frostbite, which might otherwise be caused by the freezing of moisture in the beard.

The combination of all these physical characteristics in modern mongoloids, such as the Oroqen people and some Siberian nomads, is powerful support for the argument that extremely cold conditions caused the distinctive features of the mongoloids to be selected and developed in the first place. The intensity of that adaptation is underlined by the fact that even when mongoloids began to spread south into the tropics, and their bodies became smaller and more lightly structured under the new conditions, they still retained their hair form, their cranial shape and their eye-fold.

Beyond the taiga lies the vast emptiness of the tundra, stretching north to the shores of the Arctic Ocean and east to the Bering Sea. The tundra is a treeless zone, although it does not entirely lack tree species. There are some, including willows and birches, but they have become so stunted by the fierce winds and driving snows of winter that they never grow more than waist-high. Their rate of growth is also extremely slow—a few centimetres a decade, at the most.

Most plants huddle close to the ground in a virtually continuous mat. The crowded leaves absorb the weak sunlight, warming the air trapped beneath them sufficiently to prevent their stems from freezing. Some plants have thick or furry leaves, to prevent loss of moisture in the cold but arid winds.

For most months of the year the plants of the tundra are buried beneath snow and ice. Even when the snow melts and the brief Arctic summer begins, the plants are faced with an extremely short growing season. Consequently, they have to make their new growth, burst into flower, and produce their fruits or seeds, all in less than twelve weeks. Many plants reproduce asexually, without seeds. Those with creeping rootstocks or underground stems send up shoots which become new plants. Some throw runners across the ground to put down new roots. Those that grow from bulbs produce smaller bulbs. Any seeds that are produced are encased in tough, frost-proof skins and can survive long periods of freezing, until suitable conditions bring them to life.[1]

One of the key factors in the struggle for survival by all forms of life in the northern regions of the planet is the permafrost. This is a layer of frozen ground which lies beneath about one quarter of the earth's surface: half of the USSR, more than three-quarters of Alaska, half of Canada, all of Greenland, and much of Scandinavia. Permafrost begins about twelve centimetres below the surface, and forms wherever the period of freezing air temperatures exceeds the period of summer thaw. The more intense the cold, the thicker the permafrost becomes. In parts of southern Siberia, where the mean annual air temperature is minus seven degrees Celsius, the permafrost is sixteen metres thick. Further north, where the mean annual temperature falls to minus twelve degrees Celsius, the permafrost extends down to more than 400 metres. In Verkhoyansk in Siberia, where the temperature occasionally drops to minus seventy-one degrees Celsius, the permafrost has been measured, by drilling, down to more than 1550 metres.

Permafrost lies beneath all the Siberian tundra, and yet in summer this is difficult to believe. The landscape is carpeted with vegetation, and bright with flowers. There is water everywhere, a vast maze of swamps and lakes. The reason is that when the summer sun thaws the snow and surface ice the underlying frozen layer prevents the water from draining away, and so it lies everywhere—until the winter returns to freeze it once again.

But despite appearances the tundra is, in fact, a frigid desert. Less water is precipitated here than in the Sahara, but the permafrost keeps it all at the surface. And because evaporation rates are low the land never dries out. The 'desert' therefore supports a remarkable pyramid of life. Hundreds of plant species survive in areas that receive only 150 millimetres of water a year—and most of that as snow, when the plants are dormant. But when the plants burst into brief life, vast swarms of black flies, mosquitoes, bees and other insects appear. They in turn attract huge numbers of migratory birds, which breed on the lakes and marshes of the tundra. The vegetation also supports a surprising variety of small mammals, larger grazing animals, and their predators.

It was this biomass that drew people out of the forests in the ice ages and on into the north. The few archaeological sites that have been found in

Even when the snow melts in the short Arctic summer the water lies on the surface, because of the underlying permafrost. *The permafrost layer of perpetually frozen earth extends under most of Siberia, Alaska, northern Canada and Greenland.*

Archaeological excavations in Siberia are made extremely difficult by the permafrost, because only a thin layer of thawed earth can be removed each day. *This site, at Deering Yuriak on the Lena River, has yielded some of the earliest evidence of human occupation in Siberia.*

Siberia indicate that some groups had penetrated the tundra by at least 40 000 years ago. Such sites have been less productive than they might have been elsewhere, because the deposits containing possible evidence of past human activity are frozen into the permafrost layer, and are virtually impossible to excavate. Soviet archaeologists can only operate during the brief summer, and even then they must let nature do much of the work. They clear an area, then let the sun thaw the surface of the frozen earth, down a few centimetres. The material is scraped up and put through sieves, while the workers wait for the next few centimetres to thaw. It is a tedious routine, but at least the frozen ground guarantees good preservation of anything that is found.

So far the evidence shows that when humans first ventured on to the tundra there were some spectacular animals to be hunted—especially the herds of large hairy elephants, or mammoths, which roamed widely, and the woolly rhinoceroses on the southern edge of the tundra. Each one of these animals represented a very large supply of meat, while the hides provided shelters and the bones made tools. During the long winters the hunters also began to develop skills in carving ornaments from the tusks of the mammoths. Large numbers of these giant animals were preserved in ice where they died thousands of years ago, either by drowning in bogs or falling into crevices in the frozen ground. In many places in Siberia the remains of mammoths and rhinos have been found washing out of the permafrost in which they became entombed, and over the past few centuries many tonnes of mammoth ivory have been extracted from river banks and thawed earth.[2]

There were also moose and musk oxen grazing along the edge of the forests fringing the tundra, and many smaller animals on the tundra itself: marmots, ground squirrels, rabbits, foxes and lemmings (small rodents). They, too, were affected by the permafrost, even during the summer. The lack of drainage caused by the permafrost helped plants to flourish but was a hazard for small furred animals. Lemmings, shrews and voles, which searched for food in the matted roots, were liable to wet their fur, which could be fatal for them. The permafrost also made life impossible for most burrowing animals (which is why such burrowers as moles and gophers are not found in permafrost areas).

Winter on the tundra was a testing time for the animals, just as it was for the early people. A few large mammals, such as deer, musk oxen, wolves, foxes and hares produced sufficient metabolic heat to survive the extreme cold. The smaller mammals could not stay warm in prolonged cold. Their small body mass in relation to their body surface meant rapid heat loss, and because their fur had to be relatively short and light to permit rapid movements it was a poor insulator. Their only means of survival was to seek shelter in nests and elaborate burrows, and rely on the unique insulation properties of snow. Beneath only twenty-five centimetres of snow the temperature varied very little, regardless of outside conditions. Even when the air temperature fell to minus forty-five degrees Celsius the temperature under the snow remained around zero. In this quiet, dark environment small mammals such as lemmings, weasels, shrews and voles were warmed by the summer heat still trapped in the earth.

Most of the birds that visit the tundra do so in summer, to breed, but some also overwinter there. They include the ptarmigan and the snowy owl, which have long, dense feathers and a layer of fat beneath their skin. The ptarmigan also digs tunnels into snow banks to escape the cold, and sometimes remains in these burrows for days.

Large mammals also have insulating coats and thick layers of fat. The musk ox, in particular, has an extraordinary two-layer covering, consisting of a mat of thick, soft wool, shielded by an immense cloak of guard hairs more than thirty centimetres long. It has a very compact body on stubby legs, with a thick neck and short ears and tail, and although it is more than two metres long it stands only one metre high.

There is, however, one animal above all which made possible the first movement of people into the tundra and their survival there, right down to the present. That is the large deer which in Asia and Europe is called the reindeer, and in North America the caribou. The reindeer, like the musk ox, has layers of fat under the skin, and a double coat: an inner layer of soft, curly hair, and then an outer layer of coarse guard hairs. These hairs are thicker at the end than at the base, so that when lying close together they form a virtually airtight mat. The animal is thus surrounded by a continuous layer of trapped warm air, which provides extra insulation.

The reindeer also has an ingenious circulatory system in its legs, called countercurrent exchange, to prevent heat loss through its large feet when walking on snow and ice. The arteries carrying warm blood down the legs are enmeshed with veins carrying cold blood back to the heart, so that the incoming cold blood is heated and the outgoing warm blood is cooled. By the time it reaches the feet the blood which left the heart at about thirty-eight degrees Celsius is down to nine degrees Celsius. With so little difference between the temperature of the feet and the environment there is practically no heat loss.

The permafrost sometimes gives up the skeletons of extinct Siberian animals, like this giant rhinoceros. *The exhibit is in the archaeological museum in Yakutsk.*

One of the major routes of migration northward from China during the ice ages was the Pacific coastline, with its abundant wildlife. *At least six species of whales live in these waters, and when they come in to sheltered bays to breed they are vulnerable to harpoons.*

Reindeer range very widely, from the shoreline of the Arctic in the summer to the taiga forest belt, far to the south, in the winter. They migrate between these two extremes in huge herds, sometimes numbering tens of thousands. In summer they graze on sedges and herbs, as well as willow shoots and new birch leaves. In winter they feed largely on lichens beneath the snow, or move into the edge of the taiga and eat grass and mushrooms.

The reindeer is superbly adapted to Arctic conditions, and as the early hunters gradually came to know its ways and its adaptations they found they could put it to scores of uses. The hunting of reindeer and the preparations of the products obtained from its carcase were perhaps the single most important tasks in the daily lives of these people.

The reindeer hunters worked chiefly by following the herds and spearing the stragglers—a practice in which they often found themselves competing with packs of wolves. Once the carcases had been skinned and the meat cut off the bones, the women took charge of the hides. These had to be carefully scraped clean and cured. The tanned skins were cut up, and different parts were put aside for different uses. The best sections were made into clothes and boots by the women. The remaining parts were used to make sleeping bags or covers.

A reindeer is a large animal, and after a successful hunt each family group would cache the excess meat and bones for later use. The bones and antlers were useful for making needles and tools—knives, adzes, chisels, saws. A curved rib with a thong tied to both ends and a straight piece of bone tipped with a sharp stone made a bow drill. Large bones were fashioned into knife-like blades for cutting snow blocks. Bone and horn had to replace stone for most tools and utensils, because stone was rarely available on the tundra, except by chance erosion from the permafrost into a stream bed.

The hunters on the tundra were not restricted to reindeer and other animals for their diet. In the brief summer the women collected berries and small fruits from the low-growing vegetation, and eggs from the nests of migratory birds. As winter approached, hunting became a more prolonged and demanding activity. Most of the smaller animals went into hiding, and by the time the snows came there were only a few rabbits or foxes left on the frozen plains. The hunters had to range over greater distances, seeking out the reindeer stragglers as they headed south towards the sheltered glades of the taiga.

The climatic conditions in Siberia during the ice ages cannot be known with any certainty, but it can be assumed that they fluctuated with the expansion and contraction of the polar ice sheets. The temperature gradients from the taiga north to the Arctic Ocean must have varied substantially, and it is possible that during the most intense glaciations around the pole the hunter gatherers on the tundra were forced back into the forests. As the ice sheets retreated and the climate warmed slightly, they would once again be able to move out on to the tundra and follow the herds of reindeer north.

There was, however, one other environment that was available to hunter gatherers in this northern region—one that must have continually led them on towards more northerly living areas. That was the coast of the Pacific Ocean.

As we saw in the way the Australian continent was almost certainly occupied, the coastline is a most favourable highway for hunter gatherers, providing a familiar environment and an inexhaustible supply of food. Such a highway stretches for an enormous distance from northern China to Bering Strait, taking in the long coastline of the Kamchatka Peninsula. Furthermore, in the transition from subtropical to subpolar waters the abundance and variety of marine life steadily increases. In the cold northern seas the richness of food resources is astounding.

Exactly how the coastal fauna was disposed in the ice age is of course impossible to say. Clearly, with lower sea levels and glacial extensions southwards, some species of animals may have been distributed rather differently from the way they are today. But there is no reason to believe that the overall diversity and abundance was significantly different, and merely to list the present profusion of food resources is to give a good indication of just how rich an environment the coast must have been.

Ocean fish include halibut and cod, which grow to more than fifty kilograms. There are huge schools of herring and candlefish, which can be caught in scoops; these species yield a useful oil as well as flesh. Around the rocky headlands there are ling, rock cod and other inshore fish, each requiring a different method of catching. Then there are a number of salmon species, which can be caught both in the sea and in rivers when they make their annual migration upstream to spawn. In the deep waters live some of the world's largest crabs. An Alaska or king crab may have legs that span a metre, and will feed several people. Sea urchins, octopuses and squids are common. Along the beaches and estuaries there are dozens of different shellfish that are easy to collect (but the middens of shells that the hunter gatherers must have produced have long been covered by the rise in the sea level at the end of the last ice age).

Sea birds provide another bounty, either as eggs or meat on the wing, and the colonies of birds along this coastline must be numbered in millions.

Some of the more impressive northern land mammals also visit the coast: bears, wolves and foxes congregate on river estuaries to catch salmon or just to scavenge. But it is the mammals of the sea that stir our imagination today, as they must have excited the hunters of the ice ages—either because of their numbers and their value as a source of protein, or perhaps just because of their fascinating behaviour.

One of the smallest marine mammals is the sea otter, although it attains a length of nearly two metres and a weight of forty-five kilograms. It also provides one of the finest known furs, thick and soft, and no doubt highly prized for the clothes of ice age people—as it has been in the more recent past. The trade in otter fur has so reduced this animal's numbers that it is impossible to say what its original range was, but it was probably common all along the northwest Pacific coast.

Seals have been overhunted, too, in modern times, but they still give an impressive indication of just how important a source of food they must have been to coastal hunters in the ice ages. The true seals number many species, each with its own habits, distribution and pattern of migration. They include the spotted, ringed, ribbon, bearded and—largest of them all—the elephant seal, which reaches nearly five metres in length and more than two tonnes in weight.

The eared seals include sea lions and fur seals, which are also very numerous. They are large and slow-moving animals, and would have provided copious and easily obtained supplies of meat, blubber and fur when they came ashore to have their pups. Their even bigger relative, the walrus, was a choice prize. Females grow to about three metres and weigh a tonne, while males reach four metres and twice that weight. Their ivory tusks also offered an added incentive to the hunters. The massed herds of walruses on ice floes near the coast would certainly have been an attraction to people making their way north, forever drawing hunting parties on.

Finally there are the free spirits of the oceans, the dolphins, and their majestic relatives, the great whales—twenty species of cetaceans in all. The smallest whales are the beluga or white species, and the strange narwhal, with its long, twisted ivory tusk emerging from the centre of its forehead. (This bizarre appendage gave rise to the legend of the unicorn. The narwhal probably uses it to stir up the sea floor in its search for crustaceans.) The cetaceans range in size up through the pilot whale and killer whale to the huge humpback, grey, bowhead and blue whales, the last being the largest animals that have ever existed on earth.

And so a picture gradually emerges of that slow but persistent northward human migration during the ice ages. We will never be sure what propelled those first groups, but presumably they were driven by some of the same factors which persuaded other people to head southwards out of Asia—local overcrowding, tribal warfare, geological upheavals, the loss of territory through fluctuations in sea level, perhaps just insatiable curiosity, and the human urge to find new places to live.

The movement north was two-pronged, by land and sea. Many different bands of people must have been involved, probably over tens of thousands of years. And it would hardly have been a steady progress, but more likely a series of advances and pauses, and possibly even retreats, as the ice ages intensified and weakened, the sea level rose and fell, and climates changed. Some inland hunting societies would have been forced back into the forests or on to the coast when the ice sheets advanced across the tundra. Coastal people must have availed themselves of the food resources up the rivers and in the inland valleys when conditions were good. And those that learned to adapt the fastest would have led the way.

What we can say, from even the meagre archaeological evidence from coastal and inland sites in Siberia, is that certainly by 40 000 years ago, and perhaps earlier, some people from the south had reached the eastern tip of Siberia, where Bering Strait now separates Asia from North America. How mongoloid these people looked we cannot say, but they had clearly made enough successful adaptations, physical and cultural, to enable them to survive amid bleak, treeless tundra, desolate rocky shores and freezing seas.

But those hardy hunter gatherers did more. They went on to master that frigid white world along the icy shores of the Bering Sea and the Arctic Ocean, the most inhospitable environment that humans have ever inhabited. And then—although they could not have known it at the time—they found not one but two vacant continents, waiting to be occupied. They found a new world.

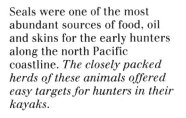

Seals were one of the most abundant sources of food, oil and skins for the early hunters along the north Pacific coastline. *The closely packed herds of these animals offered easy targets for hunters in their kayaks.*

5.

A Universe of Ice – and a New World

The area of the planet which lies within the Arctic Circle is one of the most inhuman environments it is possible to imagine. For half the year it lies frozen in the earth's shadow through the long Arctic night; for the other half the watery sun barely clears the bleak horizon. The average daily temperature never rises above ten degrees Celsius, even in midsummer, and for most of the year it plunges far below zero. The landscape is treeless, the sea shores barren and rocky, and the ocean itself black and icy. There is no timber to build shelters or make fires, and in winter the only food has to be hunted across the ice. That people should choose to leave Asia to live in such conditions, and do so successfully for tens of thousands of years, is one of the most extraordinary chapters in the story of humanity.

The Arctic region is mostly ocean, bounded by the northern shores of Eurasia and North America. Permanent ice covers two-fifths of the land surfaces, while the ocean itself is frozen over to a greater or lesser degree, according to season, by a mixture of pack ice and fast ice. Pack ice is drifting, constantly in motion. Fast ice is also floating, but attached to the land. In spring the fast ice begins to melt, and by midsummer has nearly all gone. Pack ice is thicker, and no more than half of it thaws in summer before it begins to expand again in autumn. As it grows the pack ice incorporates drifting icebergs, and finally links up with fast ice growing out from the shores.

But the Arctic Ocean never completely freezes over. There are patches of open water, polynyas, which remain ice-free through the long polar winter. Even when the air temperature falls to thirty-five degrees Celsius below zero the polynyas, although they may be dotted with ice floes, never close up. Some are little more than narrow channels in the pack ice. Others may be as large as inland seas. And they are vital elements in the story of human settlement in the Arctic, because in winter each one is a focus of life in an otherwise barren white world.

These mysterious ice-free patches of water had been known to explorers,

The first people from Asia to penetrate the Arctic Circle had to learn to live in a universe of ice and snow—the most inhospitable environment that humans have ever inhabited. *This is a glacier in the mountains of Alaska.*

89

whalers and indigenous Arctic people as good hunting grounds for a long time, but it was not until 1979 that scientists with a Norwegian expedition finally identified the mechanism which maintains them. When winter ice spreads across the Arctic Ocean, polynyas are kept open by strong winds which drive surface water away from the edge of the ice, thus causing upwelling of deeper and slightly warmer water. The rising currents bring nutrients to the surface, where they nourish microscopic phytoplankton and so-called ice algae beneath the ice floes. These support vast numbers of krill and other small shrimplike creatures, which provide food for huge shoals of Arctic cod. And this never-failing larder attracts hungry animals of all kinds, from birds to whales.

More than a dozen species of sea birds, including guillemots, fulmars, kittiwakes and gulls, gather round the polynyas in millions, feeding on the free-swimming crustaceans and small fish. Thousands of ringed seals, bearded seals and walruses also gather to feed and spend the winter. Polar bears patrol the edge of the polynyas, where the ice is unstable, and catch seals as they struggle on the heaving, slippery surface. White whales, narwhals and bowhead whales travel long distances to feed on plankton in the open polynyas. Sometimes they are deceived into entering small, temporary open patches in the ice, and are then trapped when the passage closes behind them. (Whales, being air-breathing mammals, cannot travel far under the ice.) In 1985 the Soviet ice-breaker *Moskva* broke a channel through thirty kilometres of ice off northeastern Siberia to free 3000 trapped beluga whales.

Polynyas vary in size and shape but they recur in the same places at the same time — presumably because of a combination of deep ocean currents and sea-floor topography. In winter they are oases of life in a frozen wilderness, where food resources of any kind are virtually non-existent, and early people spent many months each year beside them. It was in fact the existence of these patches of open water that made survival in the Arctic possible for humans.

Scattered thinly along the shores of the Arctic Ocean there are still many human groups who live a traditional life, or who certainly did so until well past the middle of the twentieth century. The very hostility of their chosen environment has shielded them from the impact of modern civilisation (as the desert shielded the Australian hunter gatherers). Until quite recently their lives were structured entirely around the narrow range of material things they could fashion from the resources of the Arctic, and they depended entirely upon hunting for their subsistence. From their continuing way of life it is possible to imagine what existence was like for the people who lived through the last ice age in this region, with its dramatic fluctuations in glaciation and sea levels. In particular, we can get an excellent idea of the adaptations and techniques of survival that characterised the first humans to venture into the Arctic.

The most geographically widespread of all the Arctic peoples are those speaking a group of related languages called Eskimo. They appear to have originated in northeastern Siberia, but as well as surviving along the

Chukchi Sea on the Arctic coast of Siberia they now live right across the northern fringes of North America, in Alaska, Canada (where they call themselves Inuits) and Greenland. The Aleuts and the Yupik live on the eastern side of Bering Strait and on the Aleutian Islands around the Bering Sea. There is also a whole series of people hunting across the treeless Siberian tundra and along the shores of the Arctic. The most widespread are the Yakut, and others include the Chukchi, Tungus, Evenk, Dolgan, Koryak and Yukaghir. Further along the Arctic coast, towards Scandinavia, are the Khanty and the Nenet — reindeer-herding people like the Saami or Lapps, who are spread across parts of the USSR, Finland, Sweden and Norway.

All these diverse people share a common cultural tradition which extends far into the past and is characterised by the use of similar shelters, clothes, weapons and tools, and methods of hunting and transport. Theirs is a way of life derived from the experience and adaptations of those first

All clothing worn by people in the Arctic was made from furs, chiefly reindeer. The man on the right is also wearing snow goggles, made from wood with a narrow slit. *This and other historic photographs in this chapter were taken around the turn of the century during expeditions for the American Museum of Natural History.*

colonisers of the frozen wastes of Siberia.

Because of the absence of any other suitable materials, Arctic clothing was always made from animal skins, which varied in type and number according to local circumstances. The maximum protection was obtained from having both an inner and an outer layer — the inner worn with the fur next to the skin, the outer with the fur on the outside, to shed snow. Around the camp and inside shelters the inner clothes, a tunic and trousers, were usually sufficient. The inner tunic was made of soft, fine fur, usually from reindeer, although occasionally bird skins with the downy feathers intact were sewn together for a lining. This tunic had a hood attached and long sleeves. The outer clothes, which were heavier and more restrictive, were reserved for travelling and hunting. The outer tunic also had a hood, often trimmed around the edge with wolf fur, to shield the face from driven snow. The woman's hood was usually much larger, as it could be used as a warm pouch for carrying small children. The outer trousers, which reached to just below the knee, were often made from polar bear skin. A single pelt could provide three pairs.

During the long winter months the underground houses became the centre of social life for the Koryak people. *This scene was taken in 1900 in eastern Siberia.*

For a life spent largely on ice and snow, adequate protection of the feet and hands against frostbite was particularly important. On the feet there was first a soft-skinned inner 'stocking', fur side in, which came well up the calf. Over this went a shorter 'sock', again fur side in, and with dried grass packed into the space between these two layers. Finally came the outer 'boot', fur side out, and often with a double sole. In especially cold conditions there might be another 'overshoe', making four layers of reindeer fur in all. The people protected their hands with mittens, worn fur side out, made from the finest skin from reindeer legs. The tunic was slit so that they could slide their hands inside when not using them. And despite the number of separate items an adult's outfit in total usually weighed no more than five kilograms. A modern polar exploration outfit might weigh three times that amount.

Infants carried in their mother's hood were usually naked, except for a small fur hood of their own to protect their exposed head and face. Their weight was supported by a single strap that passed under their mother's arms and fastened in front. As soon as a child could walk, and until it was five or six years old, it was fitted with a one-piece suit of very soft skin from a reindeer calf. This suit opened down the front and had a small slit at the back. As the child grew bigger it began to wear small versions of adult clothes.

Every group had its own style of clothes, and tribal markings and decorations on outfits varied widely. The basic reindeer, seal and polar bear skins could be trimmed with white fox, ermine or wolverine, and stripes or inserts to give a stylish look were popular.

As clothing varied within the traditional style, so did forms of shelter. The conical or domed tent was the typical summer dwelling right across Siberia, with a covering of sewn skins over a framework of poles. For some inland groups this same tent served as a winter dwelling, with an extra layer or two of skins. There was an opening at the top, which acted as a chimney for the interior fireplace. Wood was precious, so fire was used only in extreme cold, to take the chill out of the frozen ground. But with piles of old furs between them and the floor, the families were able to sleep naked in their sleeping bags.

In western Siberia there are remains of ice age settlements which make it clear that large animal bones, skulls and even mammoth tusks were used to weigh down the skirt of the tents and keep the fierce winter winds out. Some people piled their sleds and other spare equipment on the tent edges to help hold them down. It seems likely that this practice eventually led to the development of the sod house, which became widespread. This was a low shelter built by piling up lumps of earth. It may be that the earthen house replaced the traditional skin tent by necessity, during periods of poor hunting when skins were very scarce. The other possibility is that those remarkable innovators of the north noticed the insulating property of earth and devised earthen walls to exploit it. From there it was a short step to reduce the labour of construction and the effects of winter winds by building the earth houses in a hollow, so that they became almost subterranean. The winter blanket of snow provided extra insulation.

In eastern Siberia it was a logical step to follow the example of the animals in winter and go right underground, by excavating a pit and then covering it in. The entrance was through the roof, down a ladder consisting of a log with foot-holes cut into it. Around the Sea of Okhotsk, on the Pacific coast of Siberia, these underground houses became permanent. Many were lined with timber, and during the winter the larger ones, occupied by several families, became the centre of social life in the local group. Around the smoky central fire, while the winter blizzards raged above ground, the masked shamans related the myths, and dancers and singers acted out the sacred songs and legends. These houses were simple and ingenious responses to the Arctic climate, but the most remarkable dwelling of the polar region was the one made solely from snow and ice: the igloo.

The igloo was a significant technological achievement. It was a demonstration of the unrivalled knowledge and use of frozen materials possessed by polar people. They may have as many as a hundred words for snow, defining various gradations of consistency, colour, depth, stickiness and age. There are at least as many words to describe ice and its different states. Snow and ice—alien and potentially hostile to the bulk of humanity, except as a diversion—are vital resources to Arctic dwellers, and a familiarity with their properties could mean the difference between life and death.

By replacing, in the igloo, all conventional building materials with water in two different frozen states, the Eskimo created a masterpiece. The construction of this snow house required only a single tool: a snow knife of ivory or bone, about thirty centimetres long. The builder drew a circle on the snow between three and five metres in diameter to outline the floor area, and then cut a series of blocks of snow from the floor inside the circle, placing them round the perimeter of the circle to build up the wall. The removal of the blocks created a sunken floor. As the walls grew in height they were made to curve inwards, to form a dome. While one

In winter the Koryak people of eastern Siberia lived in underground houses, whose only access was through the roof and down the ladder made from the trunk of a pine tree. *This photograph was taken in 1900 in a village on the north shore of the Sea of Okhotsk.*

person supported the walls from the inside, another added blocks of snow from the outside. Finally a keystone block was dropped into place to complete the dome, which then became self-supporting. The gaps between the blocks were plugged with snow. Sometimes light was let in on the south or sunny side by using a piece of transparent, parchment-like tissue from the stomach of a seal, or a block of clear freshwater ice.

Inside the igloo, snow benches round the base of the wall served as tables and sleeping platforms. The entrance generally faced south, away from the polar winds. It was usually a trench roofed with snow blocks, and came up inside without cutting through the main wall and thus weakening it. The low entrance also meant that warm air inside was trapped under the dome. The shape of the igloo was perfectly designed to withstand high winds, and if snowdrifts piled up against it the only effect was to further improve the insulation.

A man and his wife working together could build an igloo for a short stay

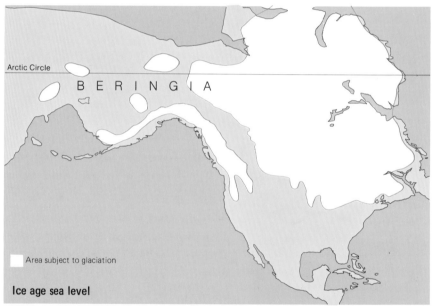

Ice age sea level

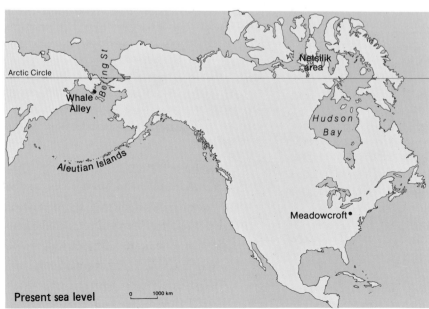

Present sea level

The igloo, being constructed of nothing but frozen water, was a technological triumph for the Eskimo, enabling them to survive months of subzero temperatures and Arctic blizzards. *The only form of lighting and heating was a lamp, which burned oil from seal or whale blubber.*

in little more than an hour. If it was located near a good fishing or sealing place, and was going to be used through the winter, it could be made larger and more permanent. The summer tent was sometimes erected inside it, to line the walls. Occasionally closely related families might build igloos with connecting openings and a communal living space.

The furnishing of the igloo was the woman's responsibility, and whenever the family moved she carried a few necessary items with her. The most important was a soapstone lamp, in the form of a shallow dish. This was placed on the snow table, with a wick made from finely chopped moss hanging over the rim. A piece of seal blubber was pounded until it began to liquefy, and put in the bowl. The lamp was lit by striking a spark from a piece of iron pyrites, or with tinder of dried moss from a fire drill, worked with a bow. Once the oily wick began to burn, the piece of blubber melted, so maintaining a constant supply of fuel. The lamp soon warmed the air in the confined space of the igloo. This caused the inner surface of the snow blocks to melt slightly, but as the water ran down the walls it froze again, forming a film of ice which further strengthened the dome.

A couple of reindeer antlers or animal bones stuck into the walls and propped up with other antlers made a rack for drying wet furs. A similar rack supported the cooking pot, suspended over the lamp with hide thongs. On the opposite side of the igloo was the sleeping platform for the family. A good layer of skins — perhaps the summer tent — was laid down over animal bones to form an insulating foundation. This not only prevented the chill of the snow from reaching the people in bed, but also stopped the heat of their bodies from melting the snow platform beneath them. On the base were spread two layers of the thickest furs available, to act as a mattress. The lower layer had the fur side down, and the upper one the fur side up. The family slept together on these — usually naked — and covered themselves with soft, warm sleeping skins, fur side down.

Thus protected from the elements with warm clothes and a house that provided a secure shelter, and was endlessly and easily renewable wher-

At some time in the past, Siberian people developed aids to transport over the snow—wooden skis, and sleds pulled by dogs. *This group of Yukaghir people was photographed in 1896 beside the Korkodon River.*

ever they went, the early tundra dwellers had the essential basis for life in the Arctic.

Their primary task, of course, was to secure their food. The Arctic, although poorly supplied with certain kinds of foods, such as plants, does have adequate resources of other kinds—if you know where and how to look. But most available food, especially in winter, is in the form of land or marine animals, the majority of which are highly mobile. Those that are not usually frequent inaccessible places, such as floating ice floes. The successful northern hunter therefore needs aids to mobility, especially on snow and ice.

Skis are an efficient means of travelling across snow, and were obviously invented a long time ago. Where they evolved is not known, but the Oroqen people of northern China use them, and at the western end of Eurasia the Lapps certainly had them several thousand years ago. For early colonisers of the polar region the chief advantage of skis would have been to increase the individual's load-carrying capacity, especially in soft snow. They would not have been much help while actively hunting, but

they would have been of great advantage to hunters keeping up with a migrating herd of reindeer, or trying to get ahead of them to set up an ambush. Snow shoes — oval wooden frames criss-crossed with hide thongs like tennis racquets, which prevented the feet from sinking into deep snow — may also have been used in Siberia. They would certainly have been more use than skis in uneven, wooded country where snow was patchy but deep in places.

The preferred means of transport in the Arctic, for both people and heavy loads, was the sled, pulled by dogs. In some places the people used reindeer, which proved easy to domesticate, but they could not match the speed of dogs or their ability to cross thin ice and deep snow. Only if no animals were available were the sleds man-hauled.

How long dogs have been used in this way in the Arctic remains unknown. The earliest dated remains of domesticated dogs, found in Iraq, go back 12000 years. There are also dates of dogs in North America nearly 10000 years old. Since the basis for all domesticated dogs is believed to have been the wolf, a semi-tame form may have been present in Siberia in the late ice age. Wolves are very common across the tundra and in the taiga, and hunters pursuing them would frequently have found litters of pups. Wolves are easily tamed when young; the early northern people may well have kept some around, feeding them on scraps of meat or seal blubber, until they were big enough to eat. And it is not hard to imagine a wolf tethered to a tent or a heavy bundle trying to drag it along in its efforts to escape. The observation of this behaviour might have suggested a use to which such animals could be put.

Whether or not such chance observations of the hauling capacity of animals stimulated the development of the sled, this elegant and remarkably efficient device became a universal means of transport in Arctic societies. The first sleds were obviously little more than a pair of skis linked by a light frame. More rigid frames and stronger, stiffer runners enhanced the sled's cargo-carrying potential, and its capacity — still unmatched by any other unpowered vehicle — to cross rough country.

The most popular and most efficient source of traction was a team of up to a dozen dogs, each one attached to the sled by a walrus or sealskin harness and a rawhide trace. Two systems of attachment were used. People living or travelling in open country or on sea ice used the fan hitch, with each dog having a trace of different length. When in full run the dogs fanned out in front of the sled, but they tended to cross one another's traces, and thus frequent halts were needed to untangle them. A more precise type of attachment was the feather hitch. The lead dog was attached by a single main trace to the front of the sled, and the other dogs were attached by short traces to the main trace, in pairs. This system allowed more precise control of the dog team and was used for travelling through broken or timbered country.

Arctic people built their sled frames from wood, when it was available, but otherwise used reindeer antlers or the bones of whales and other animals. The best runners were made from strips of baleen, the bony plates which form a plankton-sieving curtain in the huge mouth of some whales.

Even when the usual materials for sled-making were not available, the resourcefulness of the Arctic dwellers enabled them to achieve remarkable results with the most improbable materials—as they did in the construction of the igloo. In our television series we included one such example, in which a Canadian Netsilik Eskimo family demonstrated what can only be described as a masterpiece of lateral thinking.

This family had been fishing by an Arctic stream, but the sudden onset of winter forced them to pack up and move to new hunting grounds, with their skin tent and their sleeping bags, tools and spare clothes. They had walked into the area, carrying their possessions, but that had been in fine autumn weather. Now it was both snowing and freezing, and their loads, with the added weight of frozen fish, would have slowed their movements dangerously. But they had no sled, and there was no wood anywhere to make one. That, however, did not stop them.

From their tent—now frozen as stiff as wood—they took two large pieces of sealskin, each about two metres square. After folding them into rolls they lowered them into the river through a hole in the ice. This was to both soften the skins and wet them thoroughly. Meanwhile they split a few of their large frozen whole salmon into halves. Then they pulled the skins out of the water, spread them out flat, and arranged a line of overlapping frozen fish along one edge of each skin. They rolled up each skin tightly around the fish and bound it with thongs, split from reindeer skin. They laid the long, stiff rolls on the ice, and as they began to freeze they trod on them, shaping them with their feet into flattened 'boards' about twenty centimetres deep, turned up at one end. In fact they had made the runners for a sled.

They placed the runners side by side and linked them with struts of reindeer antler, lashed tightly with reindeer thongs. The result was a simple sled, quite strong and rigid, but with one severe shortcoming: the runners were rough, where the bindings cut into the skins, and would have rapidly disintegrated on rough ice. The final step, of creating hard, durable runners from what was to hand, was remarkably ingenious.

With a wooden snow shovel they dug moss from under the snow on the river bank and mixed it with fresh, light snow. In the below-zero air this mixture began to stiffen immediately, into a kind of putty. They turned the sled upside down and applied the mixture to the runners, shaping it to form a thick cushion along the running surface. Within minutes it had frozen hard, but was still rough. So they took a mouthful of water from the river and a small piece of fur, and began to polish the runners by spraying them with water from the mouth and immediately rubbing the fur along the running surface. Each spray instantly formed a new layer of ice, until there had built up a hard, strong surface, as slick as steel, and quite capable of carrying the sled and its load for a considerable distance. At any sign of wear the ice runners could be rapidly renewed.

This astonishing sequence ended with the family piling their belong-

When reindeer became domesticated in Siberia, they were also used to pull sleds. They were slower than dogs, but could move heavier loads. *These reindeer teams, bringing back meat and skins from a hunt, were photographed in the Verkhoyansk Mountains in 1902.*

ings — and their offspring — on to their new sled and setting off across the snow to a more comfortable place to spend the winter. And they had one small bonus to look forward to: when spring came and the sled runners thawed, they could eat the fish.

This is just one example of the need for flexibility in thinking that life in the Arctic imposed on people. The whole history of the human push north into Siberia and elsewhere has been one of response and adaptation to ever-changing conditions, and to a paucity of material resources (where driftwood, for example, was generally the only source of timber). The very nature of the environment ensured that flux, and not stability, was the normal state of affairs. Nothing stayed very much the same from season to season, from month to month, even from day to day. The extreme climatic variations over the year affected the way people lived to a much greater degree than anywhere else in the world. The freezing and thawing of the ocean and the rivers, the changes in day length from constant sunlight in summer to endless night in winter, the presence or absence of snow and ice — all these dictated modes of dress, styles of housing, methods of travel and, above all, techniques of hunting.

There was a pronounced seasonality about hunting, which saw the various groups of people in constant movement between the inland and the coast. The summer was, paradoxically, both the easiest and the most difficult time. The melting of the snow and ice made overland travel easier, but it also led to the scattering of the animal resources. Small game and birds could be found almost anywhere, but their day-to-day locations were unpredictable. The reindeer, after their spring migrations north into the tundra from the edge of the taiga, split into small groups and were often hard to find. There were fish in the streams, but in summer most were at sea.

Where possible, men from different families hunted in groups to maximise returns, sharing the spoils according to the role each played in the kill. This reduced the rewards for a lucky capture by a single hunter, but also lessened the individual family's chances of going hungry. With little pattern in the movement of animals, the hunters had to rely largely on accidental sightings of bears, foxes and wolves. Bears were too powerful for hunters to approach closely with spears or bows and arrows when first sighted, but often dogs were used to harass the bear and make it run. Its heavy coat caused it to overheat rapidly and become exhausted, when it could be despatched. Although killing a bear was dangerous, the amount of meat it provided was a strong incentive.

Fast, mobile animals like wolves, hares and foxes were harder to hunt in summer, but could sometimes be enticed into traps, snares and pits. Tower traps were successful with foxes. These were hollow piles of stones with a log forming a ramp to the open top, and a bait of some kind on the floor inside. The hungry fox jumped down inside to get the bait, but was unable to jump back out again.

So the summer saw little groups of people constantly on the move in a restless search for game, days of plenty alternating with days of fasting. Another problem was that in summer people had to do without the convenient and effective method of preserving surplus food that was available for nine months of the year—freezing. In the warmer months anything that could not be eaten immediately had to be dried in the sun, but because rain and fog are common in the Arctic summer this method did not always work. Only one animal product obtained in summer was habitually stored, and that was the blubber of seals and other sea mammals. Sea hunting in summer was not easy, because the animals were scattered, but if any seals or walruses were caught their blubber was wrapped in sealskins and kept for use as fuel during the long winter.

The coming of autumn saw the landscape fade silently under the first dusting of snow, and the rivers and lakes begin to freeze—but there was also a marked improvement in food resources right across the Arctic. Fish such as salmon, whitefish and char began to jostle their way up the streams to breed and were easily caught by the use of stone weirs. The hunters piled boulders across a shallow stretch of river, making dams which channelled the migrating fish into smaller, closed pens. Here, standing up to their knees in the water, the men rapidly took the fish with an ingenious spear called a leister. This has a smooth, sharp point flanked by two springy prongs with inward-facing barbs. When the point penetrates a fish the pronged barbs snap into place around its body and prevent it from slipping off. The prongs can be sprung apart and the fish removed in a second. A conventionally barbed spear would take valuable time to remove, and the flesh of the fish would be damaged in the process. With leisters, a few men can take fish as fast as they enter the pens, and often land hundreds in a single day.

After the lean provisions of summer the hunters and their families enjoyed a feast of raw fresh fish; sashimi has been around a long time in the Arctic. The surplus fish were filleted with the semicircular, razor-sharp stone knife called the ulu, and packed into stone boxes built beside the stream. Stones were piled over the top to keep out foxes and wolves.

Despite their strong attachment to their dogs, Siberian people sometimes sacrificed them to protect their village against evil spirits. *These were photographed beside the Sea of Okhotsk in 1901.*

The fish froze within hours of coming out of the water, and provided a store that could be used time and time again over the following nine months.

As the days shortened, the families who had been roaming widely across the tundra slowly drifted back towards the traditional wintering places, hunting small game and fishing through holes in ice-covered lakes along the way. Sometimes they would come across a small herd of musk oxen. If they had dogs they used them to hold the herd in a small circle while they tried to drive spears or arrows through the thick, woolly hide of these heavy-set animals.

The autumn was the great season for hunting reindeer, right across Siberia and other Arctic territories. By this time the animals were forming big herds once again for their return migration south, into the sheltered, forested valleys along the edge of the taiga. After a summer's grazing the reindeer were at their fattest, and their new fur was at its most suitable for

One of the sea mammals prized by Arctic hunters was the narwhal, whose single, twisted tusk projecting from the centre of its head gave rise to the legend of the unicorn. *An Eskimo with his catch in 1902.*

For many months of the year the people of the Arctic lived in igloos on the floating ice and hunted sea mammals, like this beluga, or white whale. *These hunters were photographed on the shores of the Sea of Okhotsk in Siberia, in 1900.*

Seals and narwhals could be harpooned from kayaks, but whale hunters needed a stouter watercraft — the umiak, made from walrus skins. *This one was photographed on Little Diomede Island in Bering Strait, where they are still in use for hunting.*

101

making clothes for the winter. This was the time when northern people had to stock up with the staples on which their lives depended: meat, fat, hides and furs, sinews, horns and bones. Groups came together along the migration routes for the killing.

In some places the hunters built converging lines of stones, designed to funnel animals past the 'firing pits', in which spearmen and bowmen lay in ambush. Elsewhere, young men and boys would creep in behind the herds and howl like wolves to scatter the adults and disorientate the calves, which could be easily run down. But the most profitable way to hunt reindeer was to attack the herds at their crossing places on rivers and lakes. Here some hunters waited in kayaks—those light, manoeuvrable, skin-covered canoes—while others hid on the banks. When the herds approached the water's edge the men leaped from cover and stampeded them in. As the animals jostled and struggled in the water the hunters in their kayaks circled their vulnerable quarry and stabbed them with spears.

Whether the kayak was an invention of the coast or the inland, it became one of the most useful and widely used aids for Arctic hunting. With its light wooden frame, made up of many small sections, it was easily repaired or modified. The covering was made of sewn skins, and these were always available. The whole thing was so light that a hunter could easily carry it past river rapids or from one lake to another. But the kayak really came into its own after the herds of reindeer and other grazing animals had headed south, the other land animals had gone into hibernation beneath the snow, and people had left the lifeless land and moved out on to the frozen sea for the winter.

The Arctic winter was a testing time for the early Siberians. They had new clothes sewn from reindeer and other furs, and perhaps some food left over from their autumn hunts on the tundra. But now, for six months or more, they would have to survive by the cold, wet, dangerous pursuit of sea mammals, on and under the ice floes of the Arctic Ocean and the Bering Sea.

Little groups of families built their snow houses on the smooth surface of the newly formed fast ice covering the ocean bays and inlets, or on the drifting pack ice bordering the open, ice-free polynyas. Then they began hunting their most accessible source of food: the several species of seals that came to the polynyas to feed on crustaceans and Arctic cod. The most convenient way of catching seals was to find one of the many breathing holes these animals maintained in the ice, and wait until one pushed its nose up for air. This could mean an all-day vigil on the ice, exposed to cutting winds and blinding snow. But if the hunter was quick with his spear he could pull out a large seal, with enough meat and blubber to keep his family for a week.

Sometimes the hunters set out in their kayaks with an assortment of spears, harpoons, hooks and lines, looking for quarry along the edge of the ice. Occasionally a curious seal would come right up to a kayak, where it could be speared. If a herd of narwhals was sighted, moving slowly along on the surface, the hunters paddled after them and tried to harpoon one or more. When the harpoon struck home the narwhal usually dived, but the harpoon head did not pull out. It was held by one of the most ingenious inventions of the Arctic people: the toggling harpoon.

This device was a carefully shaped point, made of bone or antler, which was fitted to the head of a spear shaft and attached to a line held by the hunter. When the harpoon penetrated the animal's body the tip became detached from the shaft, and as the line tightened an angled spur on the harpoon tip turned it sideways, anchoring it beneath the tough skin. The hunter threw the line overboard, its end attached to an inflated seal-skin bag. The drag exerted by this float gradually exhausted the narwhal, until it was helpless and could be despatched with spears. The penetration of the spear was increased by the use of the atlatl, a device similar to the Australian woomera, which gave the thrower's arm greater leverage.

Although narwhals, beluga whales and even walruses could be harpooned from a kayak in this way, they are so big that it was impossible to retrieve their body from a kayak, or even tow it to the edge of the ice. So in time Arctic hunters came to hunt walruses and the larger whales from umiaks. These were also made with a wooden frame and a skin cover, but the frame was much stouter, the skins much heavier, and the boat itself much larger, capable of carrying up to twelve hunters. Umiaks could only be built where there was suitable timber, and the Pacific coast of northeastern Siberia is one place where they are still used.

Umiak crews paddled long distances from shore to hunt walruses, which gathered for safety in packs on floating ice floes. Although often concealed by fogs, they gave away their location by their constant bellowing and belching. Sometimes the sea was rough, or the walruses particularly aroused by the approach of the umiak, and one or more hunters would have to go behind them, land on the ice floe, and try to harpoon or spear one. If the harpoon lodged firmly, the huge mammal usually plunged into the sea. The hunters would follow it in their umiak at a safe distance; a large walrus can grow to three metres and weigh two tonnes, and enraged and wounded animals have been known to attack boats. When the walrus was finally dead they used inflated sealskins as floats to tow the body in.

The pinnacle of Arctic hunting was whaling—the most rewarding but also the most hazardous pursuit. Harpooning whales from skin boats in icy seas was extremely dangerous, but hunters went out from many parts of the Siberian coast, into the Sea of Okhotsk and the Bering Sea in the north Pacific, and into the Chukchi Sea in the Arctic Ocean. A whale provided huge amounts of many desirable products, especially a highly prized type of skin blubber called muktuk, rich in vitamin C, and oil for lamps. Whale ribs were used as rafters in sunken huts, and as ribs in tents. Jawbones made shelters, and the tail flukes contained useful tendons. The strips of baleen in the mouth were transformed into sled runners, lashings, fishing lines and snares for catching birds and small animals.

Very little is known about the prehistoric origins of whaling in the north Pacific, but on the desolate coast of Siberia, on the western side of Bering

Strait, there is a remarkable archaeological site which clearly indicates the almost mystical significance of whales and whaling to those early people.

The site, now called Whale Alley, was found only in 1976 by a group of Soviet anthropologists while visiting a small, deserted rocky island.[1] There they found an amazing ritual complex of whale remains. More than a hundred bowhead whale skulls are set out in two great parallel rows, running along beside a beach for more than half a kilometre. Dozens of huge whale jawbones have been stuck upright in the ground along the same alignment, as if marking out a ceremonial avenue or processional way. From the centre of this avenue another track or road runs at right angles up a hill, ending in a ring of large boulders. Nearby there are 120 funnel-shaped stone-lined pits, some with layers of frozen meat and blubber still preserved in the bottom. Along the shore there are other pits, and a ring of stones where a tent village once stood.

This monumental assembly of whale bones—which has no equal anywhere else in the Arctic—appears to be some kind of religious or cult centre that linked whales and their hunters. There is at present no clue to the people who hunted the whales and placed so many of their skulls and bones on display in such a dramatic way. They appear to have no connection with the present whale hunters on this coast, nor with the nomadic reindeer herders of the hinterland. According to the few radiocarbon dates obtained so far, Whale Alley may have been abandoned only a few hundred years ago. But clearly it is a clue to the grand scale of prehistoric whale hunting, and to the adventurous nature of the early hunters who came north to these cold and desolate shores so many thousands of years ago.

Prehistoric Siberian whale hunters sometimes used the skull and rib bones of whales to make extraordinary arrangements, which had ritual significance. *Many displays like this were found by Soviet archaeologists in 1976, at a site on the shores of the Bering Sea that they named Whale Alley.*

During the last ice ages the sea level was low enough to expose a land bridge between Asia and North America, where Bering Strait now exists. *This is Bering Strait in summer, seen from the Alaskan side.*

Some time during the last series of ice ages those early Siberians were to make a momentous discovery, although they could not have realised it at the time. At some stage in their wanderings across the tundra in the wake of the herds, or in their forays along the coast in skin boats after whales or walruses, or simply in their never-ending search for a sheltered place to spend the winter, they entered an area that humans had never seen before. They crossed from Asia into North America.

This movement from one continent to another was one of the most significant migrations that humans were to make in the Pacific basin — as far-reaching in its consequences as the great southward migrations during the ice ages which discovered New Guinea, the islands of Melanesia, and Australia. And yet this dramatic human achievement is one of the most contentious issues in archaeology today — especially among modern Americans — because there is bitter disagreement about when that crossing was made, who those people were, and how they did it. Recently, however, a whole series of new lines of evidence has become available, and that evidence increasingly suggests answers to those three questions that go against many long-held convictions in this field.[2]

Earlier, we looked at the rise and fall of ice age sea levels in Asia and Southeast Asia, the consequent losses and gains in habitable land areas, and the dramatic influence that these processes had on the movement of human populations. It has long been known that similar physical processes took place in the north Pacific, but it is now becoming clear that they had equally dramatic effects on the movements of human populations. We now know that the great southward movement of modern people out of Asia some 50000 or more years ago was matched by a movement to the north.

What is new in this whole debate are the implications of the recent Australian research on the Huon Peninsula in Papua New Guinea and its fossil coral reefs.[3] With the precise dating of the reefs, and their correlation with the known rate of uplift of the peninsula (confirmed by satellite measurement), we can now be certain that for all but the last 10 000 years of the past 80 000 years the Pacific Ocean was at least twenty-five metres lower than its present level. Furthermore, for most of that time it was more than fifty metres lower. This would have been enough to expose the shallow floor of Bering Strait and link Asia to North America by a broad plain. Even when the sea was only twenty-five metres lower than at present, Bering Strait would have been reduced to a narrow, shallow passage perhaps only a few kilometres wide, easily crossed by raft or canoe in summer, and a simple walk across the ice in winter.

In short, for about 70 000 years — until the last ice age ended 10 000 years ago — Asia and North America were to all intents and purposes joined by a region we now call Beringia. For much of that time Beringia was effectively an extension of the Siberian tundra, with the same vegetation and, very likely, the same herds of grazing animals moving across it in summer, and the same populations of marine mammals living along its coastline. At its eastern end Beringia met the mountain ranges of Alaska, with sheltered valleys which would have provided shelter for the grazing herds in winter. It seems likely, therefore, that during the ice ages animals of various kinds were moving backwards and forwards across Beringia between Asia and North America. And we know that at some stage they were followed by the first human migrants.

It is the approximate date when the first people crossed Beringia that is the problem. One major difficulty is that the most recent sea-level rise — the highest for perhaps 100 000 years — has covered the area where archaeological evidence of their passing might be expected to be found.

The great ice sheets and glaciers which existed in North America during the ice ages may have been an impassable barrier, in many parts, to migrants from Siberia, but there was always a way past them — the sea. *This glacier in Alaska ends at the edge of the northern Pacific.*

Another difficulty arises from the effects on the North American continent of the last and most severe of all the cold phases of the ice ages, which reached its peak about 20 000 years ago.

During that cold phase there was a great increase in the size of the Arctic ice cap, and vast ice sheets spread across Canada and North America. It has long been thought that one huge mass of ice, many kilometres thick, was centred on Hudson Bay, and another on the mountains of northwest Canada, extending into parts of Alaska. This latter ice sheet was considered to have formed an impassable barrier to any movement of people from Asia into North America. This belief was strengthened by the fact that the oldest accepted archaeological human remains in North America were less than 14 000 years old.

The evidence, to many authorities, was conclusive: people could not have got into North America until after the peak of the ice age, when the ice cap began to shrink, the glaciers began to retreat, and an ice-free corridor opened up through Canada into the prairies of North America. There were, it is true, a few tantalising signs of a much earlier human presence in the Americas, but unarguable proof was lacking. Certainly there were no definite dates in North America that would have confirmed the arrival of people before the end of the last great glaciation.

Recent studies now suggest, however, that the North American ice sheets may have been neither as thick nor as extensive as previously thought.[4] The build-up of ice, while still formidable, may have reached its maximum in different areas at slightly different times. This means that even if the ice did cover as much of North America as was formerly thought, not all areas would have been uninhabitable at the same time. By moving from one area to another, when local conditions worsened, people may have been able to survive through the glaciation, and perhaps even continue to move southwards. Another important new conclusion is that some areas of the northwest coast of North America may never have been under glacial ice at any time.

The implications of this new view of the ice age, when considered with recent and more direct evidence of earlier human presence in the Americas, are profound. Surprisingly, these new archaeological discoveries have been taking place much further south than expected. A series of sites in South America now make it clear beyond reasonable doubt that some people must have crossed from Asia into the Americas more than 30 000 years ago, before the great glaciation reached its peak. We shall deal with these finds in more detail in a later chapter, but the main results are quite conclusive.

In northeastern Brazil a series of rock shelter sites, some with art in them, have been excavated and their contents carbon dated. They show a continuing human presence from about 32 000 years ago down to 17 000 years ago. The sites contain stone tools and fire pits or hearths, but unfortunately no human remains have yet been found. Across on the other side of South America, close to the coast of southern Chile, there is a site with fire hearths and stone tools that has been dated to about 33 000 years ago. There is a cave in the Andes that was used by humans 20 000 years ago, and two cave sites in Mexico of about the same age. Although dates of comparable age have not been found in most parts of North America, there is one cave site in Pennsylvania, Meadowcroft, which contained stone tools and bits of man-made artefacts dated by their discoverers (although not without dispute) to nearly 20 000 years ago.[5]

Putting all this new evidence together, it is clear that people had entered South America before 30 000 years ago. Since it is not in question that the first Americans came in from northeast Asia, and since it must have taken several thousand years to reach, say, southern Chile, we must recognise a span of at least 40 000 years for the existence of aboriginal Americans, or Amerinds. (The situation is very similar to that in Australia, where the oldest sites are also in the south, and the span of known aboriginal development virtually the same.) On this basis, it would seem inevitable that the very first movements across Beringia from Asia into North America must have begun more than 50 000 years ago.

As to who the first Americans were, there is virtually no dispute. While there are few, if any, human remains of ice age antiquity from either North or South American sites — other than a few teeth and bone fragments — it is quite certain that all modern Amerinds from both continents are Asian in origin. Most of them do not have the extreme flattening of the face and nose of modern Asians, but these features must have developed in Asia after the Americas were first settled. All Amerindian groups do, however, have a high incidence of shovel-shaped incisor teeth (which, as we saw in Chapter 1, is an unmistakable characteristic of Asian people). A high proportion of them also have three-rooted lower first molars, which is another Asian marker. Some Amerindian groups show a pronounced mongoloid eye-fold, while others do not — but this, again, had not fully developed in Asians by the time the first people crossed into the Americas.

Taking all these features into account, Amerinds in both continents can be divided into three major groups. The largest group consists of all of the peoples of South America, and many of those in North America, and they are the least like modern Asians. The second group is to be found in the northwest of North America, from just south of the Canadian border up through western Canada into Alaska. These Amerinds are distinct from the first group, and more like the modern people of eastern Asia and Siberia. The third group are the Eskimos of Greenland, Canada and Alaska, and the Aleuts of the Alaska Peninsula and the Aleutian Islands. They are similar to people now living on the Asian side of Bering Strait. Being the most recent migrants to the Americas they show the closest physical affinities with east Asian people. Recent studies of Amerindian languages show a similar tripartite pattern, and support the argument that all the Amerindian people came originally from Asia, and that they came in three major movements or waves.

The final question, how they came, is still wide open, and full of intriguing possibilities. Most specialists in this field, being archaeologists, and looking for sites they can excavate, tend to think of those first migrants

Many characteristic features of Siberian people were to be seen later in North Americans, and it is now generally agreed that all the American Indian peoples had their origins in Asia. *This hunter was photographed in eastern Siberia earlier this century.*

as overland explorers, waiting for a land bridge to open up to get them dry-shod across from Asia into North America, and then perhaps waiting again for an ice-free corridor to open up down to the prairies.

We think, however, that the analogy of Southeast Asia is also relevant to the north Pacific. If Asian people had a maritime technology 50000 years ago that could take them out into the western Pacific to colonise the Philippines, Melanesia and Greater Australia—all across quite considerable water gaps—then it is likely that similar technology and enterprise was employed by Asian people heading northeast, past Siberia.

Such a possibility is strengthened by the availability of a wide range of marine foods along the northern coastlines. The almost permanent closure of Bering Strait during the ice ages and the dispersal of marine mammals along the southern coast of Beringia might have further enhanced the prospects for maritime hunters moving around the northern fringes of the Pacific Ocean.

With the ice ages bringing significantly easier conditions for maritime passage up the coast of Siberia, round the Kamchatka Peninsula and past a coastline formed by the Aleutians and Beringia, and with ample food supplies all the way, it seems to us that competent mariners from Asia might have been the first people to reach North America. Such people did not need large, stable, passable land corridors or land bridges to accomplish long distance movements. Marine hunters have one great advantage: mobility. They can always move on if a chosen staging site proves unsuitable, knowing that their food resources, in effect, move with them. And in that theoretical race to enter the Americas the maritime explorers had one more incalculable advantage: the coastline is a corridor that is never closed. Even the largest glaciers or ice sheets end at the ocean's edge.

So we can perhaps imagine those early explorers, in watercraft of some kind, cautiously moving eastwards along the margins of the north Pacific, possibly bypassing Beringia, then turning southeast down the coast of North America, coastal hopping if the terrain looked too rugged, and effectively entering the new continent much further on than all the speculation about the Beringian land bridge has suggested.

As happened in Greater Australia, colonisation and settlement might well have continued down the Pacific coast, even into South America, before the expansion into the vast new continents began. However this may be, the migrations across the top of the Pacific Ocean and down its eastern shores meant that by 40000 years ago all four huge continents that form its rim were occupied by humans. And it is fascinating to reflect that those imperatives which had driven people south to the 'roaring forties' in Tasmania had also driven them to the same latitudes on the other side of the Pacific.

Meanwhile, back in Southeast Asia, at the base of that great crescent of human expansion, developments in ways of thinking about sea travel, food, and other forms of raw materials were about to vitalise the entire Pacific basin.

6.

Ten Thousand Islands

The Asian people who occupied the thousands of islands of Southeast Asia retained many mongoloid facial features, although they lost the fat-padded cheeks of the cold-adapted northern Asians. *This young woman lives in eastern Kalimantan (Borneo).*

While the earliest explorers from Asia, the vanguard of human expansion into the Pacific, were pushing down to the southern edge of Greater Australia and north through Siberia to the fringes of North America, others behind them were consolidating those conquests. In a series of migrations which may have begun as a trickle, but became a flood over the past few thousand years, people moved into the myriad islands of the Philippines, Indonesia and Melanesia (which includes New Guinea, the Solomon Islands, Vanuatu [New Hebrides], New Caledonia and Fiji). There they diversified— in appearance, language, costume, music and custom—into the most heterogeneous concentration of human societies to be found in any comparable area anywhere in the world.

Precisely how and when all those different groups of people reached their present homes, and where they came from, we will never know. We could better follow this phase of the peopling of the Pacific if we had access to human skeletal remains from this area from, say, 30 000, 20 000 and 10 000 years ago. Unfortunately, no such material has yet been found. Apart from the meagre fossils from the few cave sites mentioned in Chapter 1, there is no prehistoric skeletal evidence of the way the present diverse populations evolved. We can, however, learn a great deal from the living people and their cultures.

The region they inhabit is like no other—a vast archipelago of islands of all shapes and sizes and topographies, as if a continent had been chopped up and sprinkled across the blue tropical waters of the western Pacific. Indonesia alone can count more than 12 000 islands, and the Philippines 1400. Some islands are huge—Borneo is as big as Spain or the United Kingdom. Most of them are steep and rugged, because they are in fact the tips of mountain ranges that once emerged from the vast Sunda Plain, now submerged by the seas that rose at the end of the ice ages. Many are the exposed cones of volcanoes which continue to force themselves up through the sea floor above fault lines between jostling tectonic plates. The

111

combination of rich volcanic soils and tropical heat and rainfall—the heaviest on record anywhere—has clothed most of them in luxuriant forests, stocked with an amazing variety of plant and animal life. Today they are also the home of dense human populations, totalling more than one hundred million on the island of Java alone.

In nearly all these islands, including the Philippines and Indonesia, the majority of people are Asian in origin—mongoloids. They still display their Asian ancestry in their flat face, round head, dark eyes, yellowish-brown skin colour and straight black hair. But they have also adapted to the tropics by becoming smaller and lighter in body shape, and losing much of the fat padding that is characteristic of more northerly Asian people.

In many of the islands where mongoloids now live they replaced the original populations, a darker people known as Australoids, distantly related to Java Man. Australoids are now confined to Melanesia and Australia, and their dark colour is apparently an adaptation to a very long period of life in hot, sunny regions. The changeover of the two populations was not, of course, the result of a sudden invasion and displacement, but involved the coexistence of the two types of people until in some areas the mongoloids absorbed the Australoids. There was obviously some cultural as well as genetic mixing, and even now there remains an overlap. With

The ice-age changes in sea level created many new islands in Southeast Asia, and with them new communities dependent on the sea for their living. *This village lies off the coast of Sabah, in Kalimantan (Borneo).*

Moving east along the Indonesian island chain the people change gradually in physical type, from mongoloid in Bali to Melanesian in New Guinea. *These are people of Alor, a small island north of Timor, where the Melanesian genetic influence is strong.*

The island communities of Southeast Asia obtain virtually all their subsistence from fishing and collecting other sea food. *Fish are drying here in a village off the coast of Sabah.*

each step east from Bali along the Indonesian island chain the mongoloid character of the people decreases and the Australoid influence increases. The islanders gradually become taller and darker, and in the Aru islands, just off New Guinea, they are in fact Melanesians.

There also remain within the mongoloid populations of Southeast Asia a few odd pockets of so-called negritos. These are darker and smaller than their neighbours—the men averaging around 150 centimetres (4 ft 11 ins)—and they mostly live in mountain forests, obtaining their subsistence by hunting and collecting wild foods. There are groups of negritos in the Malay Peninsula, some of the islands of the Philippines, and in the Andaman Islands in the Indian Ocean. Their origins are obscure and puzzling, but overall they show a mixture of mongoloid and Australoid features. And that is perhaps what they are: a remnant series of hunter gatherers who were overtaken and isolated by the late ice age populations moving out of the Asian mainland. Their short stature is perhaps the result of an adaptation to forest life, just like the rainforest people of north Queensland.

Of course, all these various types of people have continually mixed and mingled in the thousands of islands dotted across the western Pacific, from the Malay Peninsula to Australia. This process was spurred by the fluctuations in sea level over the past 50 000 years. Islands grew and joined, shrank and disappeared. Populations expanded and dispersed, fused and broke up, found new ways to live as conditions altered. Families, groups, clans, whole tribes had time to develop individual artistic skills, ceremonies, music, styles of dressing. The sameness of modern Western life makes us forget the individuality of small-scale island cultures.

In many places in the Southeast Asian archipelago, groups remained small and scattered. But in others, especially by the ocean, large communities grew up, based on the abundant resources of the sea. Their markets still reflect the inexhaustible supply of food provided by the ocean. Every day the fishermen in their sailing canoes unload up to twenty different kinds of fish, as well as crabs, prawns, crayfish, clams, oysters, cockles, spider shells, squids, octopuses, jellyfish, sharks, rays and many varieties of seaweed.

All round the South China Sea, the Sulu Sea, the Celebes Sea, the Java Sea, communities put down roots beside the ocean—some, like the Bajau people of Sabah in Borneo, in the sea itself. The Bajau, or sea gypsies, live in wooden houses built on stilts far out from the coast, on the fringing reef. At high tide the ocean floods beneath the houses, bringing nutrients and all kinds of ocean creatures. At low tide the water is barely waist-deep, and the Bajau harvest the area around their houses, netting and spearing fish, trapping crabs in baskets, and wading through the shallows to collect clams and other shellfish. The reefs are so productive that the people are able not only to feed themselves with ease, but also to send valuable shipments of sea food to the shore markets. They exchange fish and shell for the few things they cannot provide for themselves: sago, tobacco and metal tools.

Groups of twenty or thirty houses form a community or village. They are virtually self-contained, and run their own affairs with little contact with or interference from the mainland authorities. There are no doctors, no schools, no government services. The Bajau are outside the system. Children born over the water grow up in canoes, and may not visit the mainland until they are teenagers.

The Bajau are known as sea gypsies because they have no ties to the land. Some of them are so attached to the sea that they never leave it. They do not even build overwater houses, but spend their entire lives in their boats, going ashore only to trade fish for sago or oil for their lamps. The typical gypsy boat is just big enough for a man and his wife, one or two children, and perhaps his mother or mother-in-law. It has a canopy or wooden roof, and woven bamboo screens make areas for sleeping and eating. In the stern there is a small kitchen, with a wok simmering on a few lumps of charcoal in a clay dish. Fishing nets are hung from one side of the roof to dry, and on the roof itself fish may be drying in the sun. On a small platform at one side there is often a chicken or two.

The sea gypsies speak their own dialect, which is incomprehensible to people on the mainland, but which links them with similar communities, sometimes great distances away. Although the Sabah sea gypsies are notionally citizens of Malaysia, other communities of the same people, speaking the same dialect, live across the Sulu Sea in the Philippines. They move freely back and forth, for trade, social contact, or just to follow the seasonal movements of the sea food which is their livelihood. The fish recognise no territorial boundaries, and nor do the sea gypsies.

In a sense, these coastal communities are still hunter gatherers, winning their livelihood from the sea. But with the exception of people like the Bajau and the forest-dwelling negrito groups, the majority of the mongoloid peoples of island Southeast Asia have changed their way of life to a dependence on agriculture. The story of that profound revolution in the way people obtain their food is the subject of our next chapter.

There remains, however, one important area in the western Pacific where we can see human societies still in transition from a hunting and gathering economy to one based on the growing of food—Melanesia. But that way of life does not put these surviving Australoids in any way behind the mongoloid populations who displaced them, for the Melanesians were in fact the leaders in one crucially important phase of the peopling of the Pacific. They were the pioneers of deep-sea ocean travel.

The heartland of Melanesia is the great bird-shaped island of New Guinea, with its spine of towering cloud-wreathed mountains, its deep slow-moving rivers, and its mantle of exuberant rainforest. Today New Guinea is separated from Australia by the shallow waters of Torres Strait, where tides race back and forth between the Arafura Sea and the Coral Sea, but for most of the last 100000 years it was part of the island continent. And as we saw earlier, the first humans were crossing the last water gaps to Greater Australia at least 50000 years ago.

While some of those first colonists were pushing south, into Australia,

The Bajau people of the Sulu Sea, between the Philippines and Kalimantan (Borneo), live in houses built over the sea, well away from land. *This is part of a community near Semporno, in Sabah.*

others were heading east, along the north coast of New Guinea. They reached the Huon Peninsula by about 45000 years ago, and then crossed the narrow strait to the large island of New Britain. It was an easy step, across a smaller gap, to the next island, New Ireland. People had colonised New Ireland by 30000 years ago, and recent research has suggested that from there they made the dangerous deep-water crossing to Buka in the Solomons by 28000 years ago.[1] At the same time, people were pushing into the interior of New Guinea itself. Some hunter gatherers even climbed into the highland valleys, where they were to establish remarkable societies that would remain unknown to the rest of the world until the 1930s.

The sequence and dates of the occupation of the other main Melanesian islands—Vanuatu (New Hebrides), New Caledonia and Fiji—remain uncertain, since no early evidence has yet been found in those groups. Considering that the first colonists had the ability to cross open ocean to the Solomons by 28000 years ago, the crossings to Vanuatu and New Caledonia do not seem much more formidable, and the recent archaeological finds in the Solomons might yet be duplicated further out into Melanesia. One explanation for the lack of evidence so far might simply be the comparatively little work that has been done in that region. It has also been suggested that early coastal sites in those islands might well have been covered by the significant rise in sea level which ended only 6000 years ago.

The Melanesians, as already stated, came from the same Australoid stock as the Australians, and this past relationship is reflected in their broad noses and lips, and generally large and projecting faces (compared to the flat faces of mongoloids). Over the past 10 000 years, however, they have been influenced genetically by other groups, including mongoloids, who have moved into the area. They have consequently developed some differences, both from Australians and within the various islands.

Melanesians are, on the whole, shorter than Australians. The men average about 160 to 165 centimetres (5 ft 3 ins to 5 ft 5 ins), while most black Australians are closer to 170 centimetres (5 ft 7 ins). And in the highlands of some islands, New Guinea particularly, there are very short people, averaging between 145 and 150 centimetres (4 ft 9 ins and

The dark skin of the Melanesians is thought to be an adaptation to a long period of inhabiting hot, sunny climates. *These are people from the village of Pere, on Manus, off New Guinea.*

Some Bajau people, called sea gypsies, are so attached to the sea that they never leave it, living out their entire lives in their boats. *Sea gypsies are always on the move, following the fish, and paying little attention to international frontiers.*

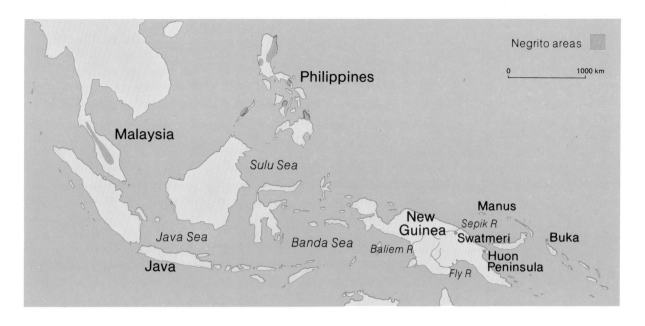

4 ft 11 ins). The reasons for this shortness of stature are not clear. It may be, as elsewhere in the world, an adaptation to high altitudes and steep slopes, because short legs reduce the energy needed to move the body over such terrain.

Despite their differences, all the islanders have some characteristics which mark them as Melanesians. Their first and most obvious feature is, of course, their very dark skin colour—hence the name of the region that they inhabit: the Black Islands. The earliest inhabitants of Melanesia were hunter gatherers, and many still are, especially those living in the dense rainforests, or by the sea. Some of the earliest archaeological sites have uncovered remains of hunting kills, which indicate that the jungle dwellers hunted marsupials, cassowaries and rodents, and gathered plant foods in the dense rainforests. Some of the early coastal sites have also produced evidence of shellfish gathering. But the early Melanesians may also have been involved in some activities not usually associated with hunter gatherers. For example, the stone axes found on the fossil coral reefs on the Huon Peninsula in New Guinea have provoked considerable debate among archaeologists.[2]

The approximate age of the axes has been established by their association with the fossil coral reefs, which have been firmly dated to 45 000 years ago. We can therefore be confident that the axes were made—perhaps in seasonal seaside camps—more than 40 000 years ago. What is even more unusual about them is their size and shape. New Guinea archaeologists have now found more than forty of them, and the biggest is more than twenty-two centimetres long and weighs 2.86 kilograms. They are made from river stones, flaked on one side, and have also been chipped away around the middle, making a waist. This groove was obviously used to bind the axe to a handle, or haft.

The Huon axes are by far the oldest hafted stone axes in the world. Similar axes have been found in the New Guinea highlands, dated to about 26 000 years ago. In Australia, waisted axes have turned up in northeast

Queensland and, surprisingly, as far south as Kangaroo Island, off the southern coastline, but their age has not been determined. It appears as though these axes may have formed part of the tool kit of the early Southeast Asians, and the knowledge of how to make them was taken to Australia and New Guinea by the Pacific's first sea-going explorers.

But this raises an intriguing question: what conceivable use would such large, heavy tools be to hunter gatherers? The most efficient tool for collecting underground foods, such as yams, is a digging stick. One obvious use for heavy axes would be for chopping down trees and clearing bush. However, if hunter gatherers in the Pacific were clearing forests in this way 40 000 years ago, they were the only people in the world doing so.

Although many Melanesians now live by horticulture and the raising of animals, there are still large groups of hunter gatherers in New Guinea and the other Melanesian islands maintaining the essential way of life of the original Australoid colonists. Some are coastal people, and obtain most of their food from the sea. Others live up the great rivers and around the lowland swamps, where a similar existence is possible. And a few, in the steep mountain forests, still live by hunting and trapping animals, and collecting plant foods. But it is the coastal and riverine people who can provide us with a window into the past, and an insight into a Melanesian tradition which made the occupation of their Pacific islands possible: their skills in boat-building and sailing. In the many different watercraft still in use in Melanesia there are clues to the history of sea travel and the ultimate conquest of the Pacific.

We have already discussed the possibility that the first people to reach Greater Australia made their sea crossings on simple but quite adequate bamboo rafts. For them to go on to reach other islands out into the Pacific would have involved a lengthy period of maritime adaptation, of experimentation with different kinds of watercraft, and sailing and steering technologies. They would have had to learn to read sea conditions and seasonal weather patterns, and the movements of the stars and planets.

It may have taken the Melanesians 20 000 years to build this maritime tradition, but it is something they have never lost. Around their islands today it is still possible to see, in the making and building of a variety of boats, some of the critical stages in the evolution of the Pacific's unique contribution to the history of ocean travel: the voyaging canoe.

If we go back even further, to Indonesia and the Philippines, there are even more basic examples of what makes this whole area a living museum of maritime technology. What is at first surprising is the discovery that, despite the availability of more sophisticated watercraft, the simple bamboo raft, which may have been used for ocean crossings more than 50 000 years ago, is still in wide use. It can be seen not only along the rivers and southern coastline of China, but in Indonesia, the Philippines, and even as far away as New Guinea and Fiji—anywhere, in fact, that suitable bamboo is available.

The offshore communities of sea gypsies run their own affairs, with little contact with the mainland. *Children born in these villages may not visit dry land until they are teenagers.*

One of the first and most enduring types of watercraft to be used in Asia and the Pacific was the dugout canoe, made by hollowing out a tree trunk. *This canoe is being made on the coast of Sabah.*

In the highlands of western New Guinea—West Irian—where deep and fast-flowing rivers are a serious barrier to travel, quickly assembled bamboo rafts are used to cross from one bank to another. Once safely across, the people usually abandon them, so easy are they to replace. On the south coast of New Guinea, particularly on the swamps and the slow-moving Fly River, bamboo rafts are everywhere in use, being easy to pole along, and capable of carrying substantial loads.

In countries where bamboo was not available, early people devised another simple watercraft: the bark canoe. No one knows where it was first discovered that a sheet of stiff, impervious material bent up into a trough and sealed at both ends would displace water and thus float, but the bark canoe appeared over a wide area of the western Pacific. This watercraft was widespread in Australia, where the bark from large eucalypts proved ideal for making canoes. In Siberia, birch bark was tough and waterproof, but lacked rigidity. The answer there was to apply a birch-bark skin to a wooden frame. The same principle was used in parts of Southeast Asia, where in the absence of suitable bark the people wove mats of thin strips of bamboo, sealed the cracks with tree gum, and used the flexible sheets to make the skin of the canoe.

It may have been the shape of the bark canoe that inspired the next form of watercraft, which was to transform water transport. This was the dugout canoe, hewn from a single tree trunk. Because of the strength and rigidity of wood, the size of a dugout was limited only by the size of the tree trunks available. The dugout must have become widespread during the ice ages, although of course there is no archaeological record of any such watercraft from those times, because of the perishability of wood.

The dugout was also the first form of watercraft that required heavy duty tools to make, and those demands may have helped to inspire the evolution of the adze. An adze is essentially an axe head rotated at right angles to the handle, and while it cannot match the cleaving or chopping function of the axe it is far superior for hollowing out and thinning wooden articles—such as a canoe. The adze is one of the most widely used tools in the world, even today, and it obviously has a very ancient origin. In the western Pacific, in the past, adzes were made from stone and, in the islands of Melanesia in particular, from the shell of the giant clam, *Tridacna*. The extremely sharp cutting edge of such a tool enabled a skilled carver to hollow out a small dugout in a single day.

Today Melanesia is one of the last strongholds of the dugout, and it may well have been here that this craft reached the peak of its development. Along the south coast of West Irian, in New Guinea, the Asmat people make some of the largest dugouts ever known. The Asmat are famous wood carvers, noted for their tall cult figures, but their canoes are even more impressive. The trees they use are tall and straight, but quite slender, so that the resulting canoe may be up to twenty metres long but no more than one metre wide. The timber is adzed to a remarkable thinness, and the prow is carved into a complex tableau of ancestral and mythic beings.

The Asmat, who live around the swampy delta of the Baliem River, are still largely nomadic, and use their great canoes to travel to new sources of food. Formerly they also carried fighting men to war with their neighbours. Even now, when ceremonies are held, special canoes are carved, painted and decorated with shells, feathers and small sculptures to provide transport for dancers, drummers and other participants.

On the north coast of New Guinea the Sepik River winds in slow, dark loops across the flat plain, like a mighty brown serpent. The Sepik is more than a thousand kilometres long, but has often changed its course, leaving oxbow lakes, lagoons and vast swamps which dry up into grasslands when the river is low. The broad, silent river is bordered by steamy lowlands of ancient silt and mud, deposited over such a long time that for much of its length there are no rock or stone formations within fifty kilometres of its banks. The Sepik has been an artery of trade and communication for tens of thousands of years, and along its low banks live some of the oldest and most traditional communities in Melanesia. Their output of carvings and pottery forms the core of New Guinea's thriving artistic life.

The village of Swatmeri, which stands beside the Sepik, is typical of both the survival in Melanesia of the hunter gatherer tradition and the use of the dugout canoe. The muddy banks are flooded each year, and crops have never thrived, so the people, who number perhaps two hundred, live by fishing. Each evening the women in their narrow dugouts go into the broad lagoons that lie next to the river, behind the village, and put out nets, suspended from floats. In the cool of dawn they paddle quietly back again and pull in their catch—a kind of plump freshwater perch. These are plentiful, and each canoe might bring back twenty or thirty fish. There are often more than the people can eat immediately, and in the tropical heat, without any means of keeping fish fresh, they would be wasted. So each day the women dry the surplus fish over smoky fires. (The fires are usually kept going under the houses day and night, to keep away the swarms of mosquitoes.)

About twice a week the men of Swatmeri load the smoked fish into their

The dugout canoe still plays an important part in the lives of people along the Sepik River in New Guinea. *Lakes near the village of Swatmeri provide a regular supply of fish.*

long, narrow dugouts and travel some way down the river to another village, a trading centre. Here they are met by people from a third village, located in the swampy forest where sago palms grow. These villagers do not fish, but collect and prepare sago, which they exchange for fish from Swatmeri. So the sago collectors have fish to eat with their sago, and the people of Swatmeri have sago to eat with their fish. It is a simple and very ancient system of living—a transitional stage beyond hunting and gathering, because neither group is entirely responsible for finding all they eat, nor are they entirely dependent upon others for their food.

There is, however, one dependency which all the people along the Sepik share, and that is on their dugout canoes. There are only two roads in to the Sepik along its entire length. For most villages and their inhabitants the only highway is the river, and the only means of transport is the dugout. There is hardly a moment when the slim, dark shapes are not gliding silently past the banks, or slipping through tree-lined channels into the still lakes.

Yet while the dugout is strong and durable, and can carry a large number of people or a heavy load of goods, it has one serious limitation: it is basically unstable. The smooth, rounded hull which makes it slip so smoothly and easily through the water offers no resistance to rocking, and even a slight disturbance of its equilibrium will tip it over. The dugout is therefore useful—and in fact only used—in smooth waters, such as slow, wide rivers or lakes.

Very early in the history of water travel this shortcoming must have been tackled, and various modifications made to the dugout canoe to make it more stable, and therefore usable in a greater range of water conditions. One improvement was to attach floats to either side of the hull, to resist rolling. A second was to extend those floats on arms, or outriggers, on either side of the canoe; this would have absolutely prevented rolling. A third possibility was, of course, to simply join two canoes side by side—to make what we now call a catamaran.

There has been a great deal of debate about which of the above solutions to the dugout problem came first, and was therefore the basis of the subsequent evolution of the characteristic Pacific watercraft. The modern distribution of these various conformations is confusing and of little help, and there is no archaeological evidence which could provide the answer.

We believe, for a number of reasons, that the initial modification of the dugout canoe was the addition of the double outrigger—that is, two arms extending from the canoe, with boat-shaped floats attached to the ends. Such floats skim the surface when the canoe is upright, but strongly resist being forced under water if the canoe tips one way or the other.

It is possible that boatmen may even have got this idea by noticing that when a pole was laid across the canoe—perhaps the pole they used for travelling along shallow waterways—it immediately provided better equilibrium, as a wire-walker's pole provides balance. The provision of fixed flotation on outriggers would have been a logical progression, and one easily accomplished by lashing a couple of poles across the top of the canoe. Now it would be possible to carry a sail—something which was out of the question with a simple dugout. And the double outrigger would keep the canoe upright no matter which side the wind was coming from.

The introduction of the double outrigger sailing canoe, whenever that may have occurred, transformed the whole pattern of inter-island travel in Southeast Asia. It was faster than any kind of bamboo raft, and in good weather it made travel across sheltered waters or reasonable sea gaps quick and easy. Even today, long after the introduction of aluminium boats and outboard motors, the small double outrigger with its swelling sail is a familiar sight around the thousands of islands in the Philippines and Indonesia. Every day the island fishermen speed out from their villages to tend their nets and empty their crab traps, knowing they can make quite a considerable round trip in their outrigger before darkness falls. Larger outriggers, some with living quarters built over the dugout hull, carry goods and passengers backwards and forwards between islands and coastal villages.

But the distribution of double outrigger craft is almost entirely restricted to the calm seas and sheltered waters of the Philippines, Indonesia and the western end of New Guinea. The reason is that the design becomes a dangerous liability in the kind of waves or swells generally met in the open sea, away from coastlines. The danger is that the craft may straddle a trough, in which case the central hull will be suspended while the two outriggers rest on the wave crests. The outriggers, not being designed to support the weight of the hull, may break off, and the hull turn over. (This happens even to modern trimarans, made from the strongest materials, for exactly the same reason.)

The double outrigger sailing canoe therefore ran up against a technological barrier in its evolution as a sea-going craft. And it was, it seems, the Melanesians, spurred by their search for islands beyond New Guinea out in the open Pacific Ocean, who found the way to break that barrier. We have no idea how they arrived at their solution, but it now seems elegantly simple: they did away with one of the two outriggers, and made the remaining one bigger and stronger. They found that it worked just as well, because even when the hull tried to roll to the open side, where the missing outrigger had been, the weight of the other outrigger acted as a counterbalance, and kept the canoe upright.

The single outrigger canoe had many advantages over the double outrigger as an ocean sailing craft. Instead of having three points in contact with the water there were only two, and this meant that the catastrophic trough-straddling overload on the outriggers could not arise. If a single outrigger found itself in this situation, with the hull on one crest and the outrigger on another, the distribution of weight and strain between hull and outrigger did not alter.

Of course, the outrigger mounting had to be strengthened to take the asymmetric load of waves striking the single float. This meant that the sides of the canoe hull had to be raised and stiffened, a modification which yielded another benefit. The building up of the sides was achieved by the

The double outrigger was an early and very successful attempt to correct the basic instability of the dugout canoe. *These craft are very widely used throughout island Southeast Asia.*

The double outrigger was liable to damage in rough seas, and the Melanesians solved this problem by removing one outrigger, thus making a much stronger craft. *Single outriggers like this one off Manus were the first ocean-going sailing canoes.*

addition of planks, lashed and pegged to the top of the dugout hull. Once invented, these side planks or strakes altered the concept of canoe-building in a fundamental way. Now the size of the hull was no longer limited by the size of the tree trunks available. The base of the hull was hollowed out from a log, as before, but the sides could now be built up, and end pieces attached with wooden pegs, to produce a much deeper and more seaworthy hull.

To this large hull a strong frame was fitted, to carry the heavy single outrigger on one side. Slots were cut in the side pieces to take the crossbars of the frame, to hold it absolutely rigid. The frame itself was made of a lattice of strong, springy timber, and the float attached to it by an ingenious linkage of palm struts and coconut fibre cords. This ensured that even when the canoe was being driven through heavy waves under sail, the outrigger float would slice through the waves, while the outrigger frame flexed up and down to absorb the shocks.

The long, slim shape of the dugout canoe makes it easy to paddle, but basically unstable. *Dugouts are best suited to calm waters, like this stretch of the Sepik River in New Guinea.*

The Melanesians also learned that, contrary to what might be thought, such a canoe sailed better with the outrigger on the side nearest the wind, where its weight provided a counterbalance to the wind pressure on the sail. (With the outrigger on the side away from the wind, the sail pressure simply forced the float and frame deeper into the waves, slowing the canoe down.) When the technique of tacking evolved, those early mariners discovered how to sail into the wind, first one way and then the other. They could tack as a sailing boat does, keeping port and starboard constant, with the outrigger alternately on the windward side and then the leeward side. They could also simply reverse the canoe at the end of each leg, sailing with first one end as the bow and then the other. This method is called shunting and keeps the outrigger constantly on the windward side.[3]

Of course, all these changes did not come about overnight. They were the product of a very long period of experimentation, the accumulation of small advances. But it does now seem likely that it was this fundamental advance in canoe-building and sailing technique which enabled the Melanesians to become the first people to venture into the open Pacific, and to occupy their island domain.

The possession of genuine ocean-going sailing craft, made entirely from trees growing on the islands, transformed Melanesian societies, by opening up trade networks with other island communities over a vast area of the western Pacific. The seaworthiness of the canoes, and the navigational skills which evolved, meant that the Melanesians could go virtually any-where they wanted to. They built sleeping and cooking facilities on their canoes, and with adequate supplies of food on board whole families or groups of people could make voyages lasting weeks or even months.

The Melanesians added living quarters to their single-outrigger canoes and carried out wide-ranging trading voyages between islands. *This canoe comes from the village of Pere, on Manus, off New Guinea.*

The evolution of the ocean-going sailing canoe brought about many other changes in the way people lived. We would like to describe one example, which we saw near Manus, a large island in the Admiralty Group, some 300 kilometres off the north coast of New Guinea.

Manus has a long history of human occupation, perhaps even dating back towards the early period of Melanesian expansion from New Guinea. The people exist mainly by collecting sago from the swampy forests and growing taro and yams. To this long-established population there was added, two or three hundred years ago, a few small groups of fishing people who arrived from other islands—probably as the result of over-crowding or tribal warfare. The newcomers were not able to obtain land from the Manus islanders, but were permitted to settle in half a dozen places along the beach, in villages like the one called Pere. It consists of a row of palm-thatched huts built over the water and just above the high-tide mark, inhabited by perhaps two hundred people. (Pere was the village where Margaret Mead lived while doing research for her book, *Growing Up in New Guinea.*)

The people of Pere have remained essentially hunter gatherers, because they live mainly on fish, crustaceans and shellfish from the lagoon in front of their village. This lagoon is about four or five hundred metres wide, and protected by a coral reef from the huge ocean swells which roll ceaselessly over the horizon. In this sheltered expanse, at low tide, the people still use an ancient technique for fishing that goes back to the traditions of the inshore and estuarine waters of Southeast Asia.

The people of Pere cannot afford large nets, but they use an ingenious alternative. From canoes they pay out long strings of palm fronds tied end to end. These float with the fronds hanging down, making a waving curtain. Men wading in the lagoon arrange the curtain in a huge circle, and then begin to slowly close the ring, pushing the screen through the water in front of them. The palm frond curtain would not of course prevent fish from swimming under or even through it, but its moving shape in the water, and its shadow on the lagoon floor, is enough to herd the fish towards the centre of the ring. Here a couple of men wait with nets mounted on a wooden frame. When the circle of men and boys is about to close completely, the men with the nets lower them to the bottom, then quickly lift them up as the ring closes amid shouts and splashes. Fish that look like escaping the nets are speared.

Such an operation, which involves many people and takes most of the day, always brings in some fish—but not generally very many. The lagoon has been well fished for generations, and each area must be rested for some days after a catch, until fish move back into it. The people of Pere— and the other fishing communities along the coast of Manus and some nearby islands—might just be able to survive on this return, but they would have to go without many other foods which they can obtain by selling or bartering fish to the inland people of Manus. Fortunately for them, their possession of ocean-going outrigger canoes opens up an entirely separate and even more valuable food resource: the pelagic fish of

The Melanesians were the first people of Southeast Asia to push out into the Pacific and settle islands across wide sea gaps. *People from the village of Pere, on Manus, taking part in a wedding celebration.*

the open ocean, such as tuna.

A tuna fishing expedition begins in the same way as lagoon fishing, only this time the target is bait fish. Watch is kept on the clear lagoons around the islands, until a large, round, tell-tale dark shadow is seen on the dazzling white sand floor. This is the shadow cast by a shoal of tiny sardine-like fish—tens of thousands of them swimming close together for safety against predators. Now a ring of men moves slowly towards the shoal of fish, walking on the lagoon floor neck-deep in the clear water. Carefully, so as not to frighten the shoal, two or three large round nets with fine mesh are slid into the water, inside the closing ring. Finally, at a signal, the men opposite the nets rush forward, shouting and slapping the water with sticks. Within seconds the centre of the circle boils up into a silver fountain of small fish, leaping frantically out of the sea. As the nets are brought together their meshes bulge with the gleaming, wriggling haul.

Waiting at the entrance to the lagoon are the ocean-going outrigger canoes, each with a large, woven, barrel-shaped basket tied to its hull, sunk in the water until only the open top is showing. Amid much shouting and laughing, the men bring their loaded nets to the baskets and pour the small fish into them in a glittering stream. With their bait kept alive by the water circulating through the baskets, the canoes head out beyond the reef, over the fathomless blue depths of the open ocean.

When all is ready, small boys on each canoe begin to scatter the bait fish overboard in handfuls, while the men beat the water with short sticks and bang the sides of the canoes. For a few minutes the ocean is calm and empty of life, except for the ominous black frigatebirds which swiftly gather and hover above the canoes. Suddenly the action begins. Attracted by the commotion, and excited by the flashing bait fish all round the canoes, the tuna come bursting up from the depths like blue-black torpedoes, slashing and biting at anything that moves. Some seize the pearl-shell hooks gleaming among the bait fish, and are swung into the canoes by the men with their short lines and palm-frond poles.

The shouting of the men and boys, the flapping of the steely blue tuna against the sides of the canoes and the harsh cries of the seagulls and frigatebirds add drama to a scene of hunting that has been going on for thousands of years—but only for those people with the ability to fish the wide, deep oceans. Such resources were never available to the people who were bound by their watercraft to inshore waters, estuaries and rivers.

The sight of these canoes heading back to Manus, their sails filled and their outriggers flashing through the waves, was a powerful reminder of the mastery of the sea that the Melanesians established so long ago, and which made them the world's first true deep-water sailors. It was also a demonstration of the kinds of technological advances in watercraft that made the settlement of the western Pacific possible—and would eventually point the way to the exploration and colonisation of the whole of the Pacific Ocean.

The Melanesians who first moved out into the ocean beyond New Guinea were the pioneers of Pacific island life. *These are children of Pere, on Manus, where Margaret Mead obtained material for her book* Growing Up in New Guinea.

The people of Pere catch fish in their lagoon by the traditional inshore method that originated in the sheltered waters of Southeast Asia. *Lacking nets, the men use a curtain of palm fronds to herd the fish into a small circle, where they can be netted and speared.*

Fishermen of Manus prepare to go deepsea fishing for tuna. *The baskets contain their live bait— small fish caught in the lagoon in nets.*

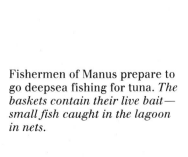

7.

Changing the Menu

To this point in the story of the human settlement of the Pacific basin, all the participants in this tremendous adventure shared one fundamental trait with the other members of the animal kingdom—they lived by hunting and gathering. The original groups of *Homo erectus* who migrated from Africa to Asia, and their successors who nearly a million years later spread out and occupied Australia, Siberia, the Americas and the islands of Southeast Asia and Melanesia—all these humans obtained their sustenance by hunting foods that moved and collecting those that did not move. It was the central preoccupation of their existence, as it had been for all branches of the human family for millions of years.

But when people everywhere began to realise that they could produce food for themselves, by growing crops, planting trees and keeping animals, they set in motion the most profound social and economic revolution that human beings have ever experienced. The domestication of plants and animals not only changed the human menu irrevocably, but it led directly to the growth of villages, towns and cities, the production of pottery and metals, the rise of industrial technology and the emergence of great civilisations—all the elements that make up modern society. And we now know that all these fundamental advances took place in the Pacific basin just as early as in other parts of the world, and quite independently of what was happening elsewhere.

The beginnings of this vital transformation in the way people lived in the Eastern Hemisphere is, however, still barely understood. The majority of the world's most important plant and animal foods were domesticated around the Pacific basin, and yet we still cannot say for sure when and where each species was first brought under human control. Much of what we do know has only come to light in the past fifteen or twenty years, but even so there have been some amazing discoveries.

There are major difficulties in trying to understand the domestication process. To begin with, it is easy to see the end product but sometimes

Rice, which was domesticated in Asia, has become the world's most important food crop, forming the staple diet of more than half of humanity. *Rice in a Torajaland market in Sulawesi (Celebes) exhibits some of its natural variations in colour.*

hard to isolate the wild ancestor, particularly where plants are concerned. Then there is the problem of deciding at what point in the domestication process the wild progenitor became the cultivated descendant. Exactly when, for example, did the jungle fowl of Southeast Asia become a domesticated chicken? There is also the question of recognising the social, economic and perhaps environmental conditions which prompted the domestication process. Finally, we need to identify in the archaeological record the physical evidence of the changes that accompanied domestication: forest clearances, field systems, irrigation works, fenced enclosures, changes in village layouts.

Such questions are difficult enough to answer, even for areas such as the Middle East, where the 'fertile crescent' has received long and detailed study. It is even more difficult to provide answers for the Pacific basin, where comparatively little work has been done. Fortunately, however, there are still societies living around the western Pacific whose lives show clearly how the idea of domestication might have occurred to early hunter gatherers, and even what they probably did about it.

One of the staple foods of Melanesia is sago, the dried starch that is obtained from the pith of the sago palm. This palm has a tall, stout trunk and large, spreading fronds, and is found in the Pacific from Thailand to Polynesia. Because of its liking for swampy conditions it flourishes wild in lowland New Guinea and nearby islands, including Manus and New Britain. The sago palm flowers only once, when it is about fifteen years old, and just before this happens the amount of starch stored in the trunk reaches its maximum. It is then ripe to be harvested. Many people of Manus obtain their protein by fishing—a form of hunting—and round out their diet by collecting and processing sago—a specialised form of gathering. We went with a group to a small offshore island to watch the affirmation of an extremely ancient tradition.

The collection of sago is a festive occasion, a day out, almost like a picnic. The men, women and children add fronds of palms and other decorative leaves to their usual costumes, and there is a great deal of talking and laughing as they pull their outrigger canoes up on the beach and prepare to make their way into the sago swamp. First they light a line of small fires at the edge of the trees, using dead palm fronds. The smoke, blown by the sea breeze, drifts through the palm trees, driving away the clouds of mosquitoes.

As the people move into the gloom of the palm grove, carrying various tools, the senior man examines the trees, looking for one of the right age. Once selected, it is rapidly felled with a steel axe. The stone axes used in the past would have been almost as efficient, because once the wire-like fibres in the outer skin of the palm trunk are severed the pithy core will not support it.

Immediately ten or a dozen men with axes and machetes swarm all over the tree, which will supply not only the raw sago but the apparatus needed to process it. The top fronds are removed, with their broad bases, and carried away to the beach, where the women begin to construct the sago-processing plant. Meanwhile, one man trims the spines of fronds into stakes, and drives them into the ground on both sides of the trunk to prevent it from rolling. At the same time a man with an axe expertly removes a broad strip of the outer skin along the full length of the trunk. Others with wooden levers prise off more skin, until the pale, creamy core is revealed. This is rich in starch, but quite hard, almost like wood, and has to be broken up and treated.

So out come the sago pounders. These are interesting tools, consisting of a handle with a narrow, curved head, resembling a pickaxe or an elongated adze. The tip can be made of stone, or in this case a renewable ring of bamboo, with its edge shaved down to make a sharp circular blade. The men line up side by side along the trunk, each with his pounder, and in unison they begin to strike the exposed pith, chanting in time with the blows. Slowly the hard, woody material is broken up, and collects in the trough which forms along the trunk where the pounders are working. It is slow, tedious work, but after an hour or two the entire core of the palm trunk is reduced to rubble. Meanwhile the women have finished making the sago-washing plant on the beach, and the processing can begin.

A load of pith is dumped into a large trough, made from the base of the palm fronds, and sea water poured over it. One woman kneads the mushy mixture, squeezing the pulp so that the starch washes out of the pith into the water. The pinkish liquid passes through a bark sieve at one end of the trough and runs down into another trough, which acts as a settling tank. As load after load of pith is washed, the sago settles out of the liquid and forms a thick, gluey white layer on the bottom of the lower trough.

At intervals one of the men lifts the sticky mass of starch out of the settling tank and drops it into another palm-frond trough, where he rolls it backwards and forwards to remove excess water. When it has the consistency of dough he pushes it into a tightly woven mesh bag, made from plant fibres. When this is hung up the weight of the sago squeezes the last of the water out through the mesh. By the end of the day there are perhaps ten firm, dry packages of sago, each weighing up to five kilograms, ready to be loaded into the canoes. The processing plant is left to return to nature.

In making these trips to collect sago, the people of Manus are maintaining the role of true gatherers of wild food. Many other people in Melanesia do the same, in places where sago palms grow conveniently close to their villages. But there are some people who have brought young palms and planted them closer to their village, for later harvesting. Others can sometimes be seen thinning out the existing groves, removing weak or damaged trees to improve the growth of the remainder. In both these activities they are manipulating the plant to suit themselves—one of the first steps towards domestication. (In Melanesia they stopped short of the next step, which was to select bigger or better palms for planting.)

In northern Australia, women from traditional communities on a food-gathering expedition will sometimes dig up a yam on the end of its vine, cut off the tuber near the stem, and push the remaining piece—still

The people of Manus, off the coast of New Guinea, still collect and process sago from the wild by breaking up the pithy heart of the sago palm and washing it. *The liquid collects in the lower trough, on the right, where the sago settles to the bottom in a white, gluey layer.*

attached to the vine—back into the ground. They know that this will produce more yams in the same place. But they do not go beyond this to plant pieces of yam in new holes. There is no doubt, however, that in many parts of Southeast Asia hunter gatherers were aware of variations in plant yields, knew the conditions in which plants grew best and even how to encourage new or better growth. Exactly when this relationship with plants became active manipulation, and eventually cultivation, is not at all clear—but recent discoveries in New Guinea suggest that in this part of the world it was surprisingly early.

When the first Europeans climbed into the highlands of New Guinea in the 1930s, a mere fifty or so years ago, they made a discovery which in its unexpectedness and its dimensions has not been equalled in modern times. After surmounting forested mountain ridges and cloud-filled passes, they came upon a totally unsuspected population of hundreds of thousands of people, living in dense concentrations in broad and fertile valleys at heights above 1300 metres. The people, while hunting in the forests, were also cultivating sweet potatoes in complex field systems. These covered the valley floors in a huge patchwork of green and brown— a sight which amazed the visitors from the thinly occupied coastal lowlands of New Guinea, with their scattered groups of hunter gatherers. Even more astonishing were the great 'sing sings' in places like the Wahgi Valley—

The first Europeans to climb up into the highlands of New Guinea in 1933 found a large and totally unsuspected population, living by intensive cultivation of yams, sweet potatoes and bananas. *These gardens near Mt Hagen are typical of the Wahgi Valley.*

gatherings of hundreds of tribesmen, decked in waving head-dresses made from the plumes of birds of paradise.[1]

The sweet potato is something of a mystery in the Pacific, because it is actually a native of South America. Somehow it was introduced into Melanesia, long before it could have been brought by European navigators, and is thought to have reached the highland valleys of New Guinea perhaps a thousand years ago. However, the highly organised system of sweet potato growing that was found in the New Guinea highlands suggested to the first anthropologists to study the area that the people there must have had a very long tradition of horticulture (although what they had grown before the sweet potato was not clear). Nor was it known when the highlanders had taken up cultivation in addition to their hunting and gathering.

Not much more was added to the understanding of the long background of food growing in the New Guinea highlands until about twenty years ago, when white settlers began to grow tea in the Wahgi Valley at a place called Kuk. The climate was ideal for tea, but many of the low-lying areas were very wet, and had to be drained with a system of ditches before planting could begin. But during the digging of the drainage channels the contractors noticed signs of much older ditches, criss-crossing the swampy areas and now filled up. Someone else had obviously tried to drain the area, a long time before. But who?

Archaeologists from the Australian National University and the University of Papua New Guinea were called in, and the results of their studies over the past twenty years have shaken many fixed ideas about when and where the process of plant domestication began.[2]

The ancient earthworks at Kuk form a very complex drainage system dating back many thousands of years. The older drains, now choked with soil and debris, extend in all directions in a tangled pattern. They show signs of digging and abandonment, followed by more digging and abandonment—phase after phase of a long and determined drainage program. They can be seen quite clearly from the air, a pattern of randomly oriented lines superimposed on the regular grid of the tea plantation. The whole complex covers an area of eight to ten square kilometres.

Among the debris in the old drains there are layers of volcanic ash from the many extinct volcanoes in the area. The evidence of frequent eruptions may explain the periodic abandonment and resumption of work on the system. Buried in the rubble-filled drains the archaeologists also found perfectly preserved long-handled wooden shovels that were used in the digging. Exactly the same tools are still in use in parts of the highlands.

The biggest ditch so far excavated was one of the principal drains for a large area. It is five metres wide and four metres deep, and has been traced for about three kilometres to its outfall, where it cuts through the side of a small hill before emptying into a river. An archaeological working party from the University of Papua New Guinea dug a trench across it for us to film. The exposed cross-section of the U-shaped ditch, now filled with

The network of ancient drainage ditches at Kuk in New Guinea is indicated by the lines running erratically across the modern tea plantations. *The earlier ditches only came to light during drainage work in preparation for planting tea here in the 1960s.*

dark mud and ash, contrasted starkly with the grey clay through which it was cut.

This drain was a major undertaking. Whoever dug it, and the network of smaller channels that feed into it, clearly understood basic hydrology. Many people, perhaps hundreds, must have been involved in the digging and maintenance of what can only have been some form of large-scale food production system.

The most surprising discovery of all, however, was that the large drain is nearly 6000 years old. Others nearby go back even further, to perhaps 9000 years ago. Since it is unlikely that Kuk was the only project of its kind, or the oldest that existed, we can safely assume that in the highlands of New Guinea some people were preparing fields for growing plants at least 10 000 years ago, before the end of the ice age. Another discovery, made much more recently, adds to the significance of these workings in the western highlands.

A few years ago two archaeologists driving through the Arona Valley, on the eastern edge of the highlands, were intrigued by a series of horizontal lines on the sides of many of the hillsides near the town of Yonki. On closer inspection, the lines proved to be grassy ridges or terraces, one above the other, a few metres apart. They had been seen before by many different experts—geologists, anthropologists, botanists—all of whom had assumed that they were natural features, either the traces of old lake shorelines, or perhaps produced by slumping of the earth during volcanic disturbances.

Excavations of a number of the formations produced a surprising result. The terraces may in fact have been natural to begin with, but some had been cut back into the slope and the base levelled, as if for the planting of crops of some kind. Such works could have only one purpose in the area around Yonki, which has a very long dry season each year. The slopes might have been terraced to collect rainfall run-off and retain the moisture in the soil. This would have permitted the growth of food plants on the terraces.

The works at Yonki have been abandoned for a very long time, and their age has not yet been fixed. But research continues, and has already established that the terraces cover an area of some hundred square kilometres. If these all turn out to be man-made, or modified in any way, they will signal the existence, some time in the past, of food-growing activities on a truly enormous scale.[3]

One major question about both Kuk and Yonki, still unanswered, is what was being grown there. The most likely candidates are taro, certain species of yams, or bananas, all of which are native to New Guinea.

More significant, perhaps, than what the people were growing is the highly organised way in which they were doing it. It points to a quite unexpected contribution to the worldwide agricultural revolution by the Melanesians, using methods which were not known in other parts of the world until many thousands of years later. If the New Guinea highlanders were clearing and preparing land for crops 10000 years ago, it can only have been because they had already domesticated and brought under cultivation a number of food plants. And that has not been shown to have occurred anywhere else before the end of the ice age.

There are major social implications here as well. At Kuk the people dug drains to get rid of excess moisture. At Yonki they created terraces to retain moisture. These vast earthworks are indisputable evidence that prehistoric people in this region were able to modify landscapes to compensate for seasonal variations in climate. To do this they had to plan large-scale engineering works and organise the labour forces needed to construct and maintain them. All this required a complex network of personal and community relationships—in fact, a society with a degree of sophistication which has previously not been recognised or even looked for in this part of the world.

With this background, it is easy to see how the highland people could have adopted the sweet potato as a staple, and developed the intensive gardening practices which so impressed the first European visitors in the 1930s. Sweet potato can grow in swampy conditions and so, presumably, it was no longer necessary to maintain the elaborate drainage systems around Kuk. As the drains went out of use and filled up, the Wahgi Valley returned to its present waterlogged condition. (The greater productivity and climatic tolerance of the sweet potato perhaps explains why it is now the dominant food crop in so many of the horticultural societies of the Pacific.)

In parts of West Irian (western New Guinea), especially in the broad Baliem Valley, swamp drainage and raised field gardens are still in use for growing sweet potato, and suggest what the Kuk system might have looked like. Using wooden spades identical to those found in the Kuk excavations, the Dani people dig drains in the swampy ground and pile up fertile mud and rotting vegetation to create gardens. They persist with this simple food production system because it continues to meet their needs.

In fact, the food grown in the Baliem Valley sustains such a large population that there is constant bickering and even fighting, particularly

The prehistoric drainage ditches at Kuk, in the Wahgi Valley in New Guinea, now filled up with debris and volcanic ash, may represent one of the earliest systems of food cultivation anywhere in the world. *The ditch cut down into the white clay, exposed here in cross-section by the trench in which Alan Thorne is standing, is 6000 years old.*

over land. In the Wahgi Valley this traditional warfare has been outlawed, and most land disputes are settled in land courts, but in the Baliem Valley the bows and arrows are still kept handy, and men sit in lookout towers to warn of impending attack. Large-scale warfare still occasionally breaks out for fundamentally the same reasons that produced so many wars in Europe over the past few hundred years: overcrowding resulting from intensive food production.

One big question yet to be answered concerns the areas where the food plants grown in the New Guinea highlands were originally domesticated. When dates ranging from 6000 to 9000 years ago were established for the Kuk drainage systems, it was at first assumed that the plants—and perhaps the knowledge of how to cultivate them—must have come from further west, in Southeast Asia. It now seems more likely, however, that New Guinea was an independent centre of plant domestication.

One ground for this suggestion is that the majority of food plants grown in New Guinea and in other parts of Melanesia—taro, yams, sweet potatoes, sugar cane, bananas—are propagated not by sowing seeds, but by simply pushing a piece of the plant or its tuber into the ground. In the continuous growing season of the tropics, the rapid regeneration of plants by this method must have been very obvious to early people. A piece of yam or sugar cane trodden into a muddy path would rapidly produce roots and shoots. It is this immediacy of results which persuades some botanists that tropical horticulture may well have been mankind's earliest method of food production, predating the sowing of seeds from cereals, which necessarily entails a rather longer interval between planting and harvesting.

Other conditions must have favoured New Guinea as a centre of plant domestication. It is a large land mass, with great environmental diversity over relatively short distances, and a profusion of native plant species. The intensity of horticulture which did eventually develop in New Guinea, at such an early date, even suggests that it may have been from this area that knowledge of domestication spread westwards, rather than vice versa.

This raises the intriguing question of why such knowledge did not spread southwards, into Australia. At the time the Kuk systems were in operation New Guinea and Australia were one continent. Even after the rising seas separated the Australians from the Melanesians there was still a great deal of contact across Torres Strait. And yet the idea of growing food did not then or later enter Australia, or if it did it was not taken up. We will return to this interesting issue later.

Meanwhile, across the huge and populous region of Southeast Asia a whole range of food plants was being domesticated by many different societies, from the jungle dwellers of Burma and the hill people of Thailand to the islanders of Indonesia and the Philippines. We may never be sure where some of those particular plants were first domesticated, or exactly when each one was brought under control. But what a list of Southeast Asian domesticates does show is that without question this region was one of the earliest and most important sources of many of the

Among the many plants domesticated in Southeast Asia are two which are now distributed universally: bananas and coconuts. *These examples are on sale in the Mt Hagen market in the highlands of New Guinea.*

most widely consumed vegetables, fruits, spices and cereals in the world today. (The only other comparably significant centre of plant domestication, as we shall see later, was the Americas.)

The list begins with several underground foods, including taro *(Colocasia)*, the giant swamp taro *(Cyrtosperma)*, and various yams *(Dioscorea)*. The breadfruit *(Artocarpus)*, native to the Philippines and New Guinea, is technically a fruit but is invariably eaten roasted, steamed or baked, and can be made into dough and preserved. One Asian bean that has achieved worldwide importance is the soybean *(Glycine)*, which was domesticated in China more than 3000 years ago. At least forty species of banana *(Musa)* are known, and their domestication is thought to have begun in Malaysia, with a few species also being brought under cultivation in New Guinea. Most varieties are quite large and are eaten cooked, as a vegetable.

The fruits of Southeast Asia are universally popular, although their origins are not always recognised. All citrus except grapefruit (which was a comparatively modern development in the West Indies) come from this region—oranges, lemons, citrons, limes, mandarins, tangerines, pomelos and kumquats. For many, the ultimate in perfume and flavour—although hardly in convenience of eating—is the mango *(Mangifera)*. There are dozens of species, and were brought into cultivation somewhere between India and New Guinea, probably more than 4000 years ago. One fruit which spread out of the region very early is the persimmon *(Diospyros)*, whose common cultivated form stems from a variety developed in Japan.

Other fruits, less well known outside Southeast Asia, are mangosteen *(Garcinia)*, jackfruit and champedak (both *Artocarpus*, and related to the breadfruit), and the large and spiky durian *(Durio)*. This has rich creamy flesh, but a powerful odour which restricts its enjoyment to elephants, tigers, orang-utans and aficionados. Another famous Asian fruit, with sweet, juicy flesh inside a spiky brown skin, is the litchi *(Nephelium)*, which is native to southern China. The closely related rambutan of Indonesia and Malaysia, which has a red skin, is thought to have originated on the Sunda Plain during the ice ages, when the Malay Peninsula, Sumatra, Java and Borneo were covered with a single tract of rainforest.

Another Southeast Asian plant which is prized for its sweetness, and now forms the basis of a huge, worldwide industry, is sugar cane *(Saccharum)*. This is a type of tropical grass, and the centre of domestication seems to have been New Guinea. It is easily grown by pushing a piece back into the ground, and was probably brought from the wild into the cultivated state by human selection for sweetness in chewing.

One fruit which has achieved universal acceptance, although more as a vegetable-like staple, is the coconut. The liquid which it contains is a refreshing drink, and the white fleshy lining and its oil are highly nutritious. The coconut is found in tropical areas all round the world, and there has been a great deal of argument about its origins. Botanists are now satisfied that this useful plant originated and was first cultivated in Southeast Asia, and has since spread both eastwards and westwards. The east coast of the

The progenitors of domesticated rice, the wild grasses in the genus *Oryza*, can still be found in wet habitats in a broad belt across Southeast Asia, from India to Vietnam. *This specimen was growing beside a small stream in northeast Thailand.*

Americas (including the West Indies) is thought to have received the coconut via Africa, while the west coast received it across the Pacific. Early voyagers in both directions out of Asia obviously found coconuts a convenient and easily transportable food resource. The nuts themselves can also germinate after several months adrift in the ocean.

The list of domesticated Southeast Asian food plants continues with spices and drugs which are now familiar everywhere: ginger, cloves, nutmeg and turmeric. Even more widely consumed, for their stimulating effect, are the leaves of one species of *Camellia*, better known as tea. The three main varieties of tea—China, Assam and Cambodia—probably sprang from a small-leaved tea which originally grew wild on the slopes of the mountains of Tibet or western China. Another stimulant which has played an important social role for many people in the Pacific is the betel nut *(Areca)*, which releases its active ingredient most satisfyingly when chewed with lime.

One of the oldest nonedible plants to be domesticated also had its origins in eastern Asia, where it may have been brought under control more than 5000 years ago. By about 3000 to 4000 years ago it was in use in Egypt, and in Europe not long after. This plant is a tall, bushy herb, and since Neolithic times the tough fibres from its stem have been used to make ropes, bindings and clothes. It was the only fibre available to the early peoples of eastern Siberia and northeastern China, where it has been under constant cultivation for at least 4500 years. In areas such as Sichuan it is grown routinely as a rotation crop with rice and oil-seed. This plant is hemp *(Cannabis)*. The narcotic effect of the resin in hemp leaves and seeds was recognised in India about 3000 years ago, but this knowledge did not reach the Mediterranean until many centuries later, and Europe as late as the eighteenth century.

Betel nuts, a widely used narcotic domesticated in Southeast Asia, are the centrepiece of a ceremony on Alor, in eastern Indonesia. *The effect of the active alkaloid in the nuts is heightened by chewing them with slaked lime.*

Cannabis (hemp) was domesticated thousands of years ago in eastern Asia for its fibres, used for making sacks and rope. Its narcotic effect was discovered much later. *In Sichuan Province in China cannabis is grown as a rotation crop with rice or wheat, and used industrially for its fibre.*

The period which began with the end of the ice ages, about 10 000 years ago, was one of great change in the Pacific basin, not only in sea levels, climates and environments, but also in human conditions. The whole vast region, from China down through Southeast Asia and out to the islands of Melanesia, saw a continuing series of revolutions in the way people lived. One of the most far reaching in its consequences was the change in the human menu, from foods that were hunted and gathered to those that were cultivated. As people everywhere began to settle down and grow their own food they laid the foundations for a new kind of society, and the rise of great cultures and civilisations. And in the role played by domesticated crops in that transformation no single plant was more important than the grass called *Oryza*, whose seeds we know as rice.

From its origins on the western shores of the Pacific, rice has become the most important food crop in the world, outranking its rivals wheat and maize. It is the staple diet of the majority of the inhabitants of the densely populated regions of the humid tropics and subtropics, and is popular everywhere else. Rice is highly nutritious, easily digested, and contains about seven per cent protein as well as valuable carbohydrates, minerals, vitamins, and even some fats. It is most nutritious with just the outer husk removed, and eaten as brown rice. Polishing the kernels to make white rice reduces its food value, but that is the way that thousands of millions of people prefer it. Altogether, rice is eaten by perhaps two-thirds of all humanity.

Botanists recognise about twenty different species of rice, including the major cultivated variety, Asian rice *(Oryza sativa)*, the next most widely grown, African rice *(O. glaberrima)*, and a number of strains that still occur in the wild. It is not known exactly when people first began to collect the seeds from wild rice and plant them in the kind of muddy, flooded margins of streams and ponds that these plants like. But from the present distribution of wild rice, and the areas of greatest diversity of the cultivated forms, it seems that domestication probably began somewhere in Asia, in a corridor which extends from northeastern India through Burma and Laos to Vietnam and southern China.

This region covers the sources of some of Asia's greatest rivers—the Brahmaputra, Irrawaddy, Mekong and Yangtze—and the ceaseless human movement along these historic routes of migration undoubtedly helped the spread of newly domesticated plants, such as rice. And although it is unlikely that the very beginning of the rice story will ever be unravelled, archaeological research is providing new insights into the profound effect of the domestication of rice on the distribution and growth of human populations in the western basin of the Pacific.

The best and earliest evidence of this comes from China. Around the lower reaches of the Yangtze and the great city of Shanghai, Chinese archaeologists have excavated several sites of early agriculture that are more than 5000 years old, and two that are even older. One of these older sites is in the village of Hemudu, near the coast east of the city of Hangzhou. It was excavated in the 1970s, after villagers had reported finding ancient relics in their rice fields.

Archaeologists from the Chekiang Provincial Museum in Hangzhou found that the village of Hemudu was built on the site of an extremely early agricultural settlement that is at least 7000 years old. The site is on the bank of a river, and for much of the time since the earlier village was occupied the ground has been waterlogged. This prevented oxygen from getting to the wooden and other perishable articles, which have been preserved in remarkable condition and now form part of a special exhibition at the lakeside museum in Hangzhou. The collection of artefacts of stone, bone, clay and wood give a picture of an early farming community, and of what the people were growing, that has not been matched anywhere else.

The excavation at Hemudu eventually extended over more than 40 000 square metres, but even then it uncovered less than ten per cent of the ancient settlement. The remains of many rectangular buildings were found, some measuring twenty metres by seven metres. These houses had stout wooden frames and were raised on wooden pilings. The timbers were put together with mortice and tenon joints, which are the earliest known examples of this woodworking technique. There were also wooden bowls coated with lacquer, made from the gum of a tree—evidence of the antiquity of this Chinese art form.

The houses contained well-made ceramic articles of many different kinds, including pots, bowls, and figures of pigs and dogs. Among the most sophisticated in design were combination clay ovens and cooking pots. The oven was a deep bowl in which a fire was lit. A cooking pot was placed over the fire, and rested on three lugs projecting from the wall of the oven. On top of the bowl went a steamer with dozens of holes in its base. It was a method of cooking food which has persisted in Chinese cuisine to this day.

In and around the houses were the bones of a number of wild animals: elephants, bears, deer, monkeys, tigers, alligators and birds. Although these indicate that the people of Hemudu still obtained part of their diet by hunting, there was convincing evidence that in fact they lived primarily by agriculture, and that their main crop was rice.

At many points in the settlement the archaeologists found large stocks of rice, the grains and husks shrivelled and blackened, but remarkably well preserved. Some of it was in pots in or under the houses, but most was in deep storage pits. Close examination of the grains shows that they are more like the modern cultivated species, *Oryza sativa*, than any of the wild species. The people had obviously been cultivating the rice in some form of

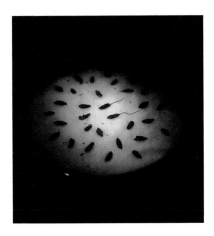

One of the earliest known areas of rice cultivation was eastern China, on the rich flood plains of the Yangtze. *These grains, excavated from the site of the 7000-year-old village of Hemudu, and now in the Provincial Museum in Hangzhou, are among the oldest specimens ever found.*

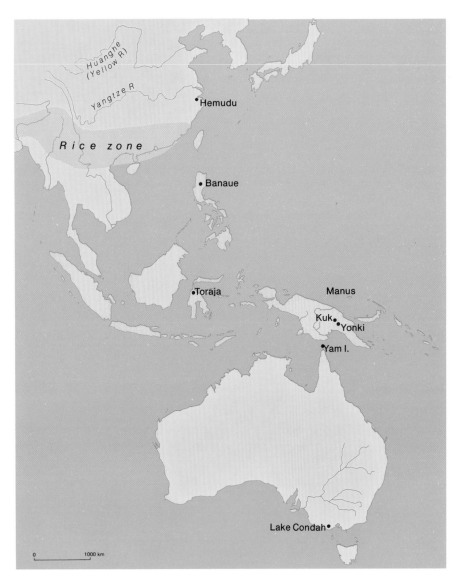

field system, and many of their tools were found. These include extremely useful hoes and spades made from animal shoulder blades, drilled and shaped and fastened to a wooden handle. At one place there was a layer, between forty and fifty centimetres thick, of a mixture of rice grains, rice husks, leaves and straw. It showed evidence of having been wet when mixed, as if the rice grains had been planted in it to grow.

It is not clear whether the people of Hemudu had already mastered the paddy method of growing rice, or just used naturally wet areas, such as the layer just described. But in any case it is clear that in this part of China, more than 7000 years ago, rice cultivation was already well established, and had led to a settled life for a sizable community. The people were still hunting wild game, but there were also bones of water buffaloes, dogs and pigs. In their shape the pig bones hint at some of the bodily changes resulting from domestication. A superb little pottery pig, seven centimetres long, confirms this impression. It has the general shape of a wild pig, but also the very clear beginnings of the heavy, sagging belly and saddle back of the familiar farm pigs in China today.

Whether or not rice cultivation in China at places like Hemudu repre-

The oldest known rice terraces in the world are at Banaue, in northern Luzon in the Philippines. *Built more than 4000 years ago, and faced with stone, the terraces are still in use.*

sents the independent domestication of rice, or is merely evidence of the spread of this knowledge, it certainly shows that by 7000 years ago rice-based agriculture was very advanced, together with all the elements of a full and settled community life.

Rice cultivation soon spread beyond the Yangtze and the Huanghe (Yellow River) into northern China, where it was apparently adapted for dry-land farming, together with two other grains already domesticated there, foxtail millet *(Setaria)* and Chinese millet *(Panicum)*. Rice was taken to the islands of Japan about 2000 years ago, where it proceeded to replace buckwheat *(Fagopyrum)*, which had been cultivated there since at least 6000 years ago. Rice also spread west into Nepal and India, and south and east into mainland Southeast Asia and all the islands of Indonesia and the Philippines.

Under the influence of the rich soils, heavy rainfall and tropical sunshine in the islands of Southeast Asia, rice demonstrated its power to influence the development of human society wherever it was introduced. A spectacular example of the stimulus that rice provided is to be seen in the steep mountain valleys around Banaue, in northern Luzon in the Philippines. The tribal people in this area have a long history, and their ancestry lies far to the west, ultimately in mainland Asia. They were originally hunter gatherers, but at some stage, when knowledge of rice cultivation reached the area, they embarked on an extraordinary program of land management and food production.

The people of Banaue cleared the forest from the steep slopes of their deep valleys, and built an astonishing series of terraces that rise in gigantic staircases up the mountain sides. Unlike rice terraces in other parts of Southeast Asia, these have retaining walls made of stone. Over the whole area, enormous tonnages of rock were quarried and put in place. The whole complex of terraces was irrigated with an elaborate system of channels and conduits, which collect water from streams high up the mountains and feed it by gravity from the highest terraces to the lowest. Bamboo pipes distribute the water wherever it is needed. And the sole purpose of this gigantic enterprise was to grow rice.

By their efforts, and with the security against famine that the high productivity of rice offered, the people of Banaue were able to transform their precipitous land into a basis for permanent survival. And although it is hard to believe when looking at these amazing terraces today, they have now sustained a sizable population for more than 4000 years. They may be the oldest continuously productive plots of earth that exist anywhere on earth.

In other areas of Southeast Asia the potential of rice can be realised with less effort. In Bali and Java the combination of high rainfall and rich volcanic soil, constantly replenished by volcanic eruptions, yields two and sometimes three rice harvests a year. No other food crop lends itself to such intensive cultivation. It is little wonder that the Southeast Asian rice bowl is the most densely populated region of the planet. In Java alone—smaller in area than the Australian state of Victoria, with its four million population—there are more than a hundred million people. Rice is everywhere in Java, and its productivity is the key to the Javanese dominance over the entire Indonesian archipelago—not just in modern times, but for thousands of years.

In the island of Sulawesi there is a quite different but no less compelling demonstration of the role that rice has played in Southeast Asian societies. There, until this century, the Toraja people were virtually isolated from the outside world in their remote mountain valleys, and their traditional life is still largely intact. The Toraja are an extremely individual people, renowned for their high, boat-shaped houses, and their practice of interring their dead in crypts high up in cliffs, guarded by life-like effigies at the tunnel entrances.

In the Toraja villages we can see what changing the human menu really meant—what it was like for people to be able to grow their own food. No longer dependent on the luck of the hunt, or collecting in the forest, they could choose when and where they obtained their subsistence. Thousands of years later, in art, music, harvest festivals, even their architecture, people like the Toraja still remember the plants and animals that made their new and settled life possible.

The Toraja, like many other Southeast Asian people, are dependent upon rice, but they go to much greater lengths than most to celebrate its planting, flowering, harvesting and husking. For the Toraja it is as if the joy and the comfort brought by the domestication of plants and animals is still fresh in their lives. They build miniature houses exactly like their own to store the rice, the golden heads tied together in fat bunches. When they draw rice from their store-houses to eat, the women join together round a wooden trough for the husking, rhythmically pounding the rice with poles,

The people who built the Banaue rice terraces in the Philippines originally came from the Asian mainland, thousands of years ago. *Banaue rice farmers still don their traditional robes and head-dresses for weddings and other ceremonies.*

In Sulawesi (Celebes) the Toraja people store their rice in special houses, built and decorated exactly like their own. *The rice keeps better if left in heads of grain, but is periodically brought out and placed in the sun to deter insect pests.*

unified by the familiar, insistent ritual.

They also decorate their houses and the rice houses with elaborate paintings of their animals in the sacred colours of black, white, red and yellow. The most honoured animals are the water buffalo and the chicken. The former is particularly highly regarded for its contribution to the cultivation of rice, by pulling wooden ploughs and rakes to prepare the paddies for planting. The only time a water buffalo is killed is for a ceremony, and then the owner mounts the head, with its spreading horns, on the front of his house. The greater the number of heads, rising one above the other on a pole like some bizarre sculpture, the greater the importance of the householder.

The water buffalo represents an important factor in the process of changing the human menu. The bringing under human control of a number of animals, alongside the domestication of food plants, was an equally crucial step in the evolution of society and the rise of civilisations.

The story of the domestication of animals in the Pacific basin is just as speculative, in its early stages, as that of the plants. Again, it is easy to see the end result, but difficult to work out how the process began. In many cases there is simply no archaeological evidence of the transition from a wild species to one that could be kept, used and bred. Some animals, like the water buffalo, have barely changed physically from their wild ancestors. Others, like some breeds of dog, have changed drastically. Pigs and chickens seem to have been domesticated in several different places, at quite different times.

There are also some puzzling anomalies in the reasons for the initial domestication of some animals. Many were obviously brought under control to provide food. Rabbits and chickens are two examples—and yet there is a strong possibility that the present importance of the chicken as a source of food may have been a secondary reason for its domestication from the jungle fowl of Southeast Asia. The clue is the continuing significance of the chicken as an element in both divination and cock-fighting. Many Asian societies still examine the entrails of a chicken or the perforations of its thigh bone as a guide to the future. Chickens are still sacrificed—not to eat, but to provide blood for ceremonies, or in the cockfighting ring in the interests of gambling.

On the other hand, the dog may have been domesticated initially in Asia as a food source, rather than as a pet or hunting aid. The largest animal ever domesticated, the elephant, has never been used for food in Asia, but only for its pulling power, its carrying capacity and, long ago, its threat as a war machine.

Many domesticates have uses beyond the obvious ones. The Asian duck and the Chinese goose are good to eat, but they also increase the productivity of paddy fields by fertilising the soil and consuming organic rubbish. Their feathers are valuable for insulating clothing in cold regions. The peacock was also domesticated for its feathers, but was generally allowed to keep them for display. The cormorant was domesticated simply for its fishing skills. Even fish were domesticated—carp to eat and goldfish

Water buffaloes are killed in Torajaland only for ceremonies, and the number of heads a man can afford to place outside his house is a measure of his importance. *The water buffalo and the chicken, two animals domesticated in Southeast Asia, are revered by the Toraja people.*

The water buffalo is undoubtedly the most important animal to be domesticated in Asia, because of the muscle power that it contributes to the intensive cultivation of rice. *Although slowly being replaced in countries like Japan by tractors, water buffaloes are still the chief motive power throughout Asia.*

in their many dazzling forms for purely aesthetic reasons.

There is no doubt, however, about the identity of the single most important animal to be domesticated in the western Pacific, and that is the water buffalo. It provides meat, skins and horns, but its greatest contribution to human society is its labour in the paddy fields. It is difficult to think of another situation where two domesticates—rice and water buffalo, plant and animal—have been harnessed so elegantly by human society. The fact that their partnership continues so widely in the western Pacific, from Japan to Indonesia, from the Philippines to India, is a measure of the potency of this ancient combination.

In some places there were local domestications. The Balinese cattle, the banteng, is one of three different wild cattle that were brought under control in Southeast Asia. It is the prime mover in the Balinese paddy fields. In fact, the contribution of animals like the banteng and the water buffalo was necessary before the full potential of rice could be realised. As living standards rise in Asia these domesticated work animals are being replaced in a few countries, such as Japan, by machines, but overall their numbers are still growing.

In the long history of domestication in Asia and the western Pacific, spread over the past 10 000 years, scores of plants and animals have been brought under human control and put to use. The energy they provided helped to build populations and stimulate migrations. Most of the hundreds of cultures and ethnic groups that make up the human kaleidoscope in the southwest Pacific are, ultimately, the products of this complex process.

And yet in all this vast sea of domestication there was, until about 200 years ago, one large and intriguing exception—Australia. The revolution in the human menu that swept through Asia (and, later, through the Americas and right across the Pacific) stopped at Torres Strait.

The reasons why Australia was the only continent to miss out on the agricultural revolution is one of the most puzzling questions in the prehistory of this continent. As we have seen, the people of New Guinea have been growing food in gardens and keeping pigs and chickens for thousands of years, but Torres Strait seems to have acted like some kind of selective cultural filter. There was considerable contact across this narrow water gap, studded with island stepping stones, and many elements of Melanesian culture were taken up by people across the north of Australia: outrigger canoes, skin drums, smoking pipes, funeral posts, and certain initiation ceremonies and hero cults. But the idea of growing food was not one of them.

There are people living on some of the islands towards the eastern end of Torres Strait who show the complex mixing of Melanesian influences, physical as well as cultural. They follow the Melanesian method of growing food by plant propagation, using cuttings instead of seeds. Yam Island is one of this group, and it is named because of the fondness of the inhabitants for a type of small, sweet yam, which the women grow in gardens. The Australian tribes in Cape York and Arnhem Land gathered the same yam, but never planted it.

The Australians had many other opportunities to take up the domestication of animals and plants. The Cape York people knew of pigs from the Torres Strait islanders, and no doubt tasted them occasionally. In Arnhem Land they learned about rice from the Macassans who came down to the north coast of Australia to catch beche de mer—sea slugs. They watched those fishermen plant tamarind trees, and soak the fruit to obtain a sweet, pungent drink. They saw trade goods like tobacco and, later on, iron axes, that stemmed from a settled agricultural life. And yet with all this contact—pressure, even—the Australians never took up the systematic planting of crops or the keeping of animals for food or work (with the possible exception of the dingo, although even that was not commonly used for food or hunting, but kept more for company).

For most modern people, who live by growing crops and raising animals, the question is, why did the Australians not want to do the same? We think it is better to ask the question the other way round: why did they prefer to remain hunter gatherers? Support for this view—that it was a matter of choice rather than some lack of enterprise or ability—comes from recent discoveries in Australia, which indicate that at least some of the first Australians were on the verge of food production and a settled life, but in a most surprising way.

The evidence comes from western Victoria, where in many of the coastal rivers large numbers of eels migrate upstream in the spring, and return to the sea in autumn. These eels, which may be a metre long and as thick as a man's arm, spend the summer in the lakes and swamps that are filled by the winter rains. They gathered in particularly large numbers in Lake Condah, a broad but shallow depression in a landscape covered with ancient lava flows, which floods each winter but dries out during the hot summer. Eels are good eating, and were a traditional food of many Victorian tribes. But the people who lived around Lake Condah did more than just fish for eels. They turned eeling into an industry.

The discovery of how they did this was made only a few years ago, by archaeologists who were puzzled by a series of what appeared to be low, man-made stone walls around Lake Condah. The people who once lived in the area have long been dispersed, but a few members have come back to run a small business near the lake, showing tourists something of their traditional life. With their help, the archaeologists finally pieced together an extraordinary story.[4]

It seems that the local people used the boulders of lava lying everywhere to build elaborate walls and networks of channels around the edge of the lake. As the water level fell during the summer the eels were directed into narrow places where they could be more easily caught in wicker funnels. The people understood the ecology of the lake system, and the behaviour of the eels. In places they dug long channels to move the eels in particular directions, and in at least one case they actually extended the range of the eels to swamps previously devoid of them.

So productive and reliable was the eel-trapping system that the people could depend upon this food supply for many months of the year, and built stone houses to live in during the fishing season. These were round, domed structures about two metres across, with a stone wall about a metre high, and a roof of branches and bark. There was a doorway in one side. Each housed a family, who lived in the house for at least six months of the year, and in some areas they may have stayed there through the winter. Around Lake Condah several hundred of these dwellings have been found. They are represented now only by their circular stone bases, since it is 150 years or more since anyone lived in them.

What this means is that beside Lake Condah there was a settlement or village of between 800 and 1000 people, tending their traps and living semi-permanently in a large community. The discovery has surprised many anthropologists, because this kind of life, very close to a settled existence, is not what has been normally associated with hunter gatherers, and certainly not with the first Australians. It shows, however, that those Australians had the capacity to organise their labour, to manage their environment, and to adopt a social way of life which elsewhere led to domestication. This example demonstrates quite clearly that they could have gone on to domestication, had they wanted to, using native plants and animals or simply by adopting other people's domesticates.

We believe that the reason the Australians did not follow this course was a simple one. They made a choice not to become farmers or herders because they preferred a way of life that maximised the time they had available for other things. We have shown earlier how relatively easy it was for them to obtain all their needs from the Australian environment. For most human groups in the other continents, domestication brought great advantages: increased productivity of food, population growth, and a more complex material existence. But with the increase in population densities came diseases, intermittent famines and wars.

More importantly, perhaps, the increasing amount of time that agricultural societies had to spend in the fields, or working at trades to pay for the products of the fields, meant the loss of a great deal of time—time that the Australian hunter gatherers put into creating their remarkable heritage

A surprising recent discovery in Victoria is that Australian aboriginal people built these rock walls as part of an elaborate eel-trapping industry. *The traps were built around Lake Condah, which flooded after winter rains and contained huge numbers of eels.*

Beside their eel traps, the people around Lake Condah in Victoria built permanent houses with stone walls, which they occupied during the eeling season. *This ring of stones, with a doorway in front, was originally a wall about a metre high, roofed with sticks.*

The persistence of the hunting and gathering tradition in modern people is reflected in fish markets, which sell what is essentially wild food. *The day's line-up of fresh tuna, ready to be sold, in Kyoto.*

of art, music, myth and natural philosophy. This was overlooked by the first European arrivals, who saw only an aimless, nonproductive society of wanderers, forever on 'walkabout'. There is nothing like the 'shiftlessness' of the nomad to arouse the scorn—and envy—of those tied to the plough or the desk.

And there is one final point to remember about domestication. In our modern view of history we tend to see hunting and gathering as that phase of human existence which preceded domestication and the agricultural revolution, as if most people gave up the one and now live by the other. It is important to realise, however, that while domestication certainly changed the lives of most of the world's people, hardly anyone now lives by domesticated food production alone. For many highly developed societies around the Pacific basin, hunting and gathering is still a crucial element in getting the food they need for their populations, and even providing export earnings to obtain other requirements.

Japan has huge fleets of ships scouring the Pacific Ocean for the whale meat, tuna, squids, crabs, crayfish, prawns and dozens of other kinds of sea food that form the bulk of the protein intake of that country's hundred million people. The fishing fleets of the United States, Canada, the Soviet Union and Chile are all engaged unceasingly in a vast hunting and gathering operation in search of food. The world's annual consumption of fish—nearly eighty million tonnes—is greater than that of beef, sheep meat, pork or chickens and eggs, and is the largest source of protein in humanity's diet.

Every fish shop and restaurant in every city is a showroom for the spirit of the hunter gatherer, as are many shelves in the local delicatessen, with their field mushrooms, truffles, seaweeds, herbs, snails, frogs' legs, wild ducks, canned and bottled oysters and mussels. The menu has certainly changed, but it is rare to find one, in any of the countries around the Pacific Ocean, that does not include something wild, to remind us of our roots.

8.
A New Cutting Edge

The change from hunting and gathering to the growing of food, and the settled life that this transition made possible, was one of the most important milestones in the long march of humanity. But there was another equally significant revolution, and it had to do with the tools that made the evolution of modern society possible.

Human beings have been making and using tools for more than a million years, as far back as *Homo erectus*. Even before then some of the ancestral primates, such as *Homo habilis* (handy man), had a rudimentary tool-making capacity. The earliest tools were probably not the ones usually thought of in terms of hunting—flint-tipped arrows, spears and clubs—but simple utensils such as digging sticks for gathering plant foods, containers for collecting shellfish, and sharp rocks for chopping tough plants and cracking open nuts and shells. The possession of the skills required to fashion such tools is sometimes regarded as the cultural marker which sets the human species off from other primates.

The first humans to leave Africa and discover the Pacific Ocean were certainly tool-makers. To judge by the little archaeological evidence that has been discovered from that period, the manufacturing skills that *Homo erectus* brought to Asia improved slowly but surely over the next million years. What began as simply worked pebbles, found in sites such as the cave of Peking Man in China, gave way to finely-made hand axes. The flaking of edges gave way to polishing. Hand tools were attached to hafts or handles. (As we have seen, this advance may have taken place in the Pacific basin earlier than anywhere else.) Closer to the end of the ice ages, we find modern people making more specialised tools from a whole range of different materials—stone, wood, bone, ivory, hides, animal teeth and shells.

We know exactly how these tools were fashioned and used, because in Australia their use continued right down into modern times, and there are still many people who know how to make them. In the desert, northwest of Alice Springs, there are extensive stone quarries where the Walbiri people

People had been making tools for millions of years from stone, bone, wood and shell before these natural materials were replaced by metals. *An elder of the Walbiri people working in a traditional flint quarry in the desert.*

have been making their tools for thousands of years. Over a wide area the red desert sand is littered with lumps of white quartz, where outcrops have been broken up and the large pieces shattered in the production of choppers, blades and points. Like the Walbiri, people in different parts of the continent knew from long experience the special qualities of the tool-making stone in their territories—flint, quartz, chert, ironstone and others—and these were traded very widely for specific uses.

Two elders of the Walbiri people took us to their quarry and showed us how they make a hunting spear. The process begins with the search for a suitable shaft. The best is a tall, slender stem of one of the very tough desert bushes. Such a stem is straight, flexible and will bend without breaking, but this means that it cannot simply be snapped off at the base; it must be cut. The Walbiri spear-maker casts around until he finds a sharp-edged rock, and with this hacks at the stem until it is severed. He removes any bends by heating the shaft over a small fire and straightening it, just as modern furniture makers shape cane or bentwood. When satisfied that it will fly straight and true, he roughly sharpens the thicker end with a hand-held stone axe. Sitting cross-legged, holding the shaft with the point towards him, he strikes towards the point with the axe, adze fashion. To finish the point he needs another tool, a scraper, which his companion has been making.

The scraper is one of the most versatile and useful tools possessed by the desert people. It has a short handle and a narrow, chisel-shaped quartz blade fitted into a notch at one end, then cemented in place with gum from the spinifex, the characteristic plant of the desert. The gum is found as a tiny grain at the base of each spinifex leaf. If enough of these are collected and warmed together over the fire, they can be kneaded into a single lump, like putty. The gum is packed and shaped around the scraper blade where it fits into its handle, and when it cools it becomes rock hard.

Using the scraper with the blade towards him, the spear-maker shaves the spear to a fine, smooth point, the wood curling off the blade as if from a steel chisel. Besides such applications, scrapers are used like an adze for hollowing out coolamons (wooden bowls), and for cutting meat and cleaning the skins of animals. The handle of the woomera or spear-thrower often has a scraper blade attached to it. For the desert Australian, it was as essential as a modern bushman's pocket-knife.

Right across the Pacific there is another material that was highly prized for tool-making. This is obsidian, which looks like black or dark green glass but is in fact a form of granite with a high silica content, ejected in molten form from active volcanoes. The rapid cooling gives obsidian a glassy structure, and it can be flaked into razor-edged blades, just like man-made glass.

Obsidian was one of the most valuable trade goods carried by the Melanesians on their inter-island voyages, and both finished artefacts and raw pieces have been found on islands many thousands of kilometres from their origins. The various deposits of obsidian have distinctive chemical fingerprints, which enable pieces to be traced to their source, even thousands of years later. The islands of the southwest Pacific have a number of well-known deposits of obsidian, created by volcanic eruptions along the 'ring of fire', which runs from New Zealand past New Guinea and through Southeast Asia. One of the most famous was on Lou Island, in the Admiralty Islands off the north coast of New Guinea.

On many volcanic islands the obsidian is scattered over the slopes of the volcanoes, glittering in the sun like broken glass. On Lou Island, however, large lumps of obsidian have become buried beneath soil and leaf litter and overgrown by jungle. To obtain their supplies the islanders dug shafts and hauled up the lumps in baskets on the end of ropes made from palm fibres. They struck cores and large flakes from the obsidian boulders for further flaking and working into pieces suitable for trading. One of the Lou Island products which was widely known throughout Melanesia was a large knife. It had a triangular blade of obsidian up to about fifteen centimetres long, cemented with tree gum into a palm-wood handle decorated with coloured lines.

Archaeological research in the western Pacific in recent years has begun to reveal the remarkable extent of the obsidian trading networks. Obsidian from Lou Island has been found nearly 3000 kilometres away, in Vanuatu (New Hebrides), and there is a recent find which hints at an obsidian trade a similar distance to the west. From another source at Talasea in New Britain, known to have been in use at least 6000 years ago, obsidian was

One of the most useful of all materials for producing sharp blades was obsidian, the natural glass ejected from volcanoes. *Obsidian blades struck from a core are as sharp as razor blades, and do not rust.*

traded to Fiji, 3700 kilometres to the east. These finds have delineated a trading area spanning something like 7000 kilometres, equivalent to the distance from London to Teheran, or from Cairo to Cape Town. It may well have been that Melanesian obsidian, carried far and wide on the great outrigger sailing canoes, was one of the world's most widely distributed prehistoric trade goods.[1]

But meanwhile, far to the west, fundamental and far-reaching changes were taking place in the field of tool-making. In different parts of the world, including Asia and Southeast Asia, people were beginning to use a new class of materials—metals.

How the discovery of metals was made we will never know, but it may simply have begun with a gradual awareness that among the materials to hand were some that behaved differently from others—substances that did not crack or chip like stone, and which could in some cases be hammered into different shapes. They may also have seemed heavier, and perhaps more interesting than other materials, because they could be polished to a particular sheen.

These substances would have been the so-called 'native' metals—not chemically linked to other elements as mineral ores, but existing in a more or less pure state, lying about on the ground or exposed in rock outcrops.

Gold is unusual among metals in that it never forms compounds with other elements, but always occurs in its pure metallic state, whether as grains or nuggets in streams, or as veins in solid rock. Gold was therefore probably one of the first native metals to be collected, being conspicuous by its incomparable lustre and its soft, glowing colour. Silver also sometimes occurs in a pure state, but is generally combined in its ores, often with lead. Copper is found in a pure metallic state in outcrops of copper ores, although it is generally oxidised by the atmosphere to an unmistakable green. The commonest metal, iron, is never found in a natural metallic state, because it has a great affinity for oxygen and rapidly oxidises or rusts away to powder. Iron from meteorites is different; it contains a high proportion of nickel, which inhibits rusting. Early people sometimes found pieces of meteoric iron and beat them into bracelets, or perhaps tips for spears or arrows. The other available metals were more suitable for ornaments, and that is almost certainly how they were first used.

The native metals available for collection on the surface of the earth were obviously in limited supply, and could have played only a minor role in tool-making. But when, perhaps only four or five thousand years ago, the early metalsmiths learned to extract metallic elements from their ores in the rocks—to smelt metals—humanity crossed an awesome threshold, from the stone age into the age of metals.

Exactly when and where this great discovery was made is the subject of a whole scientific discipline, archaeometallurgy. Understandably, most of the work in this field so far has been carried out in the Near East and the Mediterranean basin, where the terms bronze age and iron age have been adopted to describe the first significant technical phases in the rise of

Western industrial society.[2]

Until quite recently, the Eastern Hemisphere did not figure very prominently in these studies. With the exception of China, no part of the Pacific basin had demonstrated an early use of metals. The Chinese use of bronze was thought to have followed the bronze age in the West, as the result of the diffusion of knowledge eastwards, through India and central Asia. And outside China there was no evidence of any kind of early bronze age at all.

But in the past twenty years all this has changed, in a most dramatic way. From discoveries made in parts of Southeast Asia and China, it is now clear that many people on the western shores of the Pacific not only made the transition from the stone age to the age of metals quite independently of the rest of the world, but in certain critical areas of metallurgy they were ahead of everyone else. This revolution in thinking began because of discoveries in a most unlikely place: the rural landscape of northern Thailand, not far from the Mekong.

The Mekong is one of the great rivers of Asia, rising in Tibet and flowing through or bordering China, Burma, Laos, Thailand and Vietnam. It has been one of the major highways of human traffic for thousands of years, and it now seems that it has also witnessed some of the most significant developments in the history of metallurgy in Southeast Asia. Even today, along the banks of the Mekong, people still practise one of the oldest methods of gathering native metals.

The grains of gold in the Mekong River bed are too fine to be separated from the sand by panning, so a blob of mercury is placed in the dish to absorb the gold. The mercury-gold amalgam is later heated to drive off the mercury as a vapour and leave the gold behind. *The mercury is recovered by condensation.*

The villagers who live near the Mekong discovered long ago that the sands and gravels in the river bed contain finely divided particles of gold, washed down from seams of metal in the distant mountains. During the dry season, when the river is very low and the bed largely exposed, the women pan for the gold in the puddles, using large wooden dishes. When they have washed the gravel and sand out of the dish they are left with a heavy black silt containing the gold, which is too fine to be separated out by panning. So the women add a blob of mercury to the dish, and roll it around in the silt. This curious liquid metal happens to be one of the few metals that gold is attracted to, and so the mercury picks up the gold dust from the silt. The amalgam of mercury and gold is poured into a bottle and taken back to the village. When the mixture is heated in a small pot, under a clay cover, the mercury vaporises, leaving the gold behind. The mercury is collected by condensation, to be used another day.

As a technique, this simple form of gold-winning is really the hunter-gatherer end of metallurgy—merely collecting something that is lying about. The next step—recognising the different metals in their ores in rocks and then learning how to extract them—was a tremendous technological advance. It was as momentous a step forward as the domestication of plants and animals that it accompanied, and involved a stretch of the imagination, allied with intuition, which even now is difficult to explain.

It was undoubtedly noticed very early that native metals could be softened by heating, and even melted and cast. But merely heating metallic ores does not release the metal, which is linked by strong chemical bonds to elements such as oxygen, sulphur and carbon. One of the commonest metals in the ancient world was copper, which occurred in surface outcrops as copper oxide, the vivid green stone called malachite. Copper oxide was almost certainly one of the first metals to be smelted— but to do this requires two conditions: first, intense heat, at least 1084 degrees Celsius, which is the melting point of copper; and second, an atmosphere low in oxygen but rich in carbon gases. Such an atmosphere draws off the copper from its ore and reduces it to its molten metallic form. To achieve both these conditions required the invention of some form of forced draught to the furnace, and a furnace design which enabled the carbon fumes from the burning fuel to drive out the oxygen and maintain a reducing atmosphere.

The smelting process, even in its simplest form, is an exacting technology, and it is still not clear how it was conceived and carried out by people whose only industrial tools were still made of stone, clay, wood and similar natural materials. Even when the early metalworkers had learned how to make tools and weapons from copper, the progression to using bronze—an alloy of copper and tin which has great advantages over pure copper in casting, working and holding an edge—was another technical advance of daunting complexity. It was therefore natural, perhaps, for archaeometallurgists to have believed for so long that such a series of technological discoveries could have taken place only once, and that from some original centre of innovation the knowledge must have spread to other parts of the world. That centre of innovation was, it was thought, the Middle East, and especially Mesopotamia, where bronze came into use around 5000 years ago.

No other possibility seemed viable. The only use of bronze of even comparable antiquity was in China, and there the bronze age seemed to have come into flower without a copper age, which in the West had been a necessary and logical precursor of the bronze age. This suggested that the knowledge of how to make bronze must have been introduced from the West. And that was the picture of the development of metallurgy—until a discovery in a quiet village called Ban Chiang in northern Thailand.

Ban Chiang is not far from the Mekong. The villagers migrated across the river from Laos only about 200 years ago, cleared the land, and began to farm, using water buffaloes to plough their rice paddies—as they still do. They built their village on a low rise on the otherwise flat Khorat Plateau. Thereafter life in this rural backwater flowed uneventfully, until an American university student came in 1966 to carry out sociological research. One day he tripped over what looked like a tree root in a village street but which turned out to be the rim of a buff-coloured pot, with unusual designs painted on it. He took the pot back to Bangkok and showed it to an American art historian. Soon after, similar pots began to appear in the Bangkok antique stores, as villagers in Ban Chiang dug them up from under their houses and in their garden plots. The pots, with their swirling designs in red and dark grey, were unlike those of any known culture, and puzzled archaeologists in the Thai Department of Fine Arts.

The upshot was a series of excavations in and around Ban Chiang in the 1970s by teams from the Thai Department of Fine Arts and the Museum of the University of Pennsylvania in Philadelphia. This research produced results of a quite extraordinary nature.[3]

What soon became apparent, as the excavations went down right in the centre of Ban Chiang, was that the village had been built on the site of a very ancient and quite unknown culture. Beneath the streets of Ban Chiang were large numbers of burials, profusely stocked with grave goods. Dating of the pottery showed that the earliest settlement had begun at least 5000 years ago, and that the culture there had survived until about 1600 years ago, when it disappeared. The graves contained hundreds of beautifully made pots, painted in bold, distinctive designs. Although looters had robbed many of the burials, some of the finest examples are now in the museum that was built on the outskirts of Ban Chiang in 1984. There were other interesting objects in the burials, too—polished stone tools and fluted clay rollers that may have been used for printing cloth or woven mats. But what was really astonishing to the archaeologists was the discovery of a profusion of metal objects in the burials.

Here was an area with absolutely no record of any kind of prehistoric metal usage—or even a modern industrial revolution—yielding sophisticated cast objects of high-quality bronze. There were spearheads, axes, adzes, bracelets, anklets, neck bands, small bells and other ornaments of exquisite workmanship, all demonstrating excellent skills in alloying, casting, annealing and other techniques of metallurgy. Perhaps the most technically impressive object was a bimetallic spearhead, which had a forged iron blade with a cast-on bronze socket.

What sent tremors through the archaeological world were the ages of these metal objects. What they showed was that here in northern Thailand

One of the earliest forms of metal winning was the recovery of alluvial gold by panning. *These women are working in the bed of the Mekong River in Thailand during the dry season.*

The existence of a Southeast
Asian bronze age was only
confirmed in the 1970s, by the
discovery of very ancient
bronze and pottery grave goods
beneath the village of Ban
Chiang in northeast Thailand.
*Ban Chiang pots are sought by
collectors worldwide, but some
of the finest are now in the
village museum at Ban Chiang.*

there had been a fully developed bronze industry by about 4000 years ago,
and by about 3000 years ago the metalsmiths in the area were making
objects combining two different metals and two different metalworking
techniques. In another excavation not far away, at Non Nok Tha, the
archaeologists found a cast, socketed digging tool made of pure copper,
dated to about 4700 years ago (which they named WOST, for 'world's
oldest socketed tool'). There were no cultural or stylistic influences to
suggest that any of the objects had been brought in from somewhere else.
On the contrary, there was positive evidence, in the form of stone moulds
and crucibles with fragments of bronze still in place, to confirm that they
had been made locally.

The unexpected discovery of a fully fledged bronze age in Southeast
Asia, at a time not greatly different from its counterpart in the West, was
startling enough. But what fascinated archaeologists was the nature of the
artefacts found in the burials. There were no obviously warlike tools—no
battleaxes, swords or maces. Furthermore, most of the bronze objects had
been buried with children, not adults. The bronze age in Thailand seems to
have been a remarkably peaceful one, when metals were used not for
warfare but for farming, hunting and personal decoration.

One major mystery about the finds at Ban Chiang and Non Nok Tha
remained: where had the raw materials come from? Tin might have been
panned from the streams in the area, but there are no copper deposits on
the plains of northern Thailand. Eventually, a Thai government geologist
found outcrops of copper and signs of very ancient mining on the banks of
the Mekong. In 1984 a team of Thai and American archaeologists explored
and excavated the site at Phu Lon. They found an ancient copper mine—
the oldest yet discovered in Southeast Asia—and signs of ore preparation
and smelting dating from at least 3000 years ago. Phu Lon could have been
the source of the bronze objects found at Ban Chiang.[4]

The excavations in northern Thailand are among the very few attempts
to uncover the prehistory of this region. And yet their results have already

raised some intriguing questions—not only about an unsuspected bronze age in the western Pacific, but also about the wider history of metallurgy. It is clear that by about 5000 years ago some people in Southeast Asia had developed a highly sophisticated bronze industry, with a level of technology at least as advanced as any in the West. This bronze culture had apparently emerged from a village or rural setting and not, as elsewhere, from urban, militaristic societies. Here, bronze and iron were for tools and ornaments, not weapons. The ancient people of northern Thailand preferred to remember their children rather than mark the power and prestige of adults. It was a peaceful bronze age, without parallel anywhere else in the world.

And these discoveries raised one more intriguing possibility, directly related to the bronze age in the West. Bronze needs tin—about ten per cent for the best results—but although enormous amounts of bronze were made in the Middle East and the Mediterranean the source of the tin has never been identified. Tin is usually recovered as alluvial grains from streams, but it has to come initially from veins in rocks—and no sufficiently large-scale deposits of tin have ever been found in the region. It was presumed to have been imported, but the most relevant mentions in the ancient records refer only to a trade in tin from somewhere far to the east. A few years ago Soviet geologists announced the discovery of tin deposits in Afghanistan, but it is not known whether they were mined in bronze age times. That distant source, it now appears, may well have been Southeast Asia. Some of the world's largest tin deposits are to be found in Thailand and the Malay Peninsula. Furthermore, these deposits are close to large resources of copper.

It is possible, therefore, that Southeast Asia may not only have been the source of the tin for the Western bronze age, but perhaps also the source of the knowledge of how to make bronze. The major copper deposits in the Mediterranean—in Cyprus, Anatolia and Spain—are not at all associated with tin. In Southeast Asia the two metals are found relatively close together, and in some cases tin is actually a contaminant of the copper ores. This would clearly enhance the possibility of an accidental discovery of the metallurgical superiority of a copper-tin mixture over pure copper.

Whether or not future research adds weight to this suggestion, it already appears beyond doubt that there was an independent discovery and development of metallurgy in Southeast Asia. Very little is known about the further development of a tropical bronze age, because the only significant archaeological finds so far have been those in Thailand. But it is logical to assume that trade in metals formed an important element in the large-scale cultures which began to evolve.

As we have seen, the domestication of plants and animals provided the basis for a settled life in Asia, with the growth of large communities and complex states. The discovery of metals enhanced those societies, as the mining and possession of mineral resources—a durable asset—helped to produce wealthy ruling classes, which in turn supported artists and craft workers. In Southeast Asia these burgeoning civilisations built up vast

The bronze objects found at Ban Chiang in Thailand show considerable skills in metal-working and casting, on a par with those of bronze age metal-workers in the Middle East. *This hunting spear had the tip symbolically bent over before being included in a burial.*

maritime trading networks, exchanging spices, timber, and ideas, as well as the desirable new products in bronze.

Part of the great export of culture and trade goods out of mainland Southeast Asia in the period between 2000 and 3000 years ago were some unusual metal artefacts that have attracted to themselves a considerable mystique. These are bronze Dong Son drums—so named because of the examples that were excavated in 1924 at Dong Son in northern Vietnam. The typical Dong Son drum has a squat, round base, with a single tympanum or sounding plate on top. It has decorative motifs cast into the top and sides, representing scenes of people in procession, or sometimes in boats. Round the rim there may be small sculptured figures of animals. When being played, lugs around the sides enabled the drum to be suspended by ropes—although it could also be carried under the drummer's arm and played that way. The entire drum was made by lost-wax casting from quite good quality bronze (although some examples contain lead as well as copper and tin, in order to make the metal flow more easily into the moulds).[5]

For centuries these drums have been turning up all over Southeast Asia, from southern China in the north to Indonesia in the south, and from Burma in the west to the fringes of New Guinea in the east. They have become collector's items, and as well as accumulating in museums they keep appearing in antique shops in places like Hong Kong, Bangkok and Jakarta. And yet despite the interest in them, and the detailed studies made of them, it has proved impossible to say where or when they were first made, or even who made them. Some are obviously very old, but some are clearly more recent, and there is a possibility that the tradition of making them was continued in a number of places, including Burma.

The most likely origin of the Dong Son drums seems to be the area of northern Vietnam and Yunnan Province in southern China, where they formed part of a growing bronze production of socketed axes, spearheads, sickles, fish-hooks and ornaments. These metal goods were widely traded throughout Southeast Asia, where they were greatly prized. The drums

The earliest copper mine in Southeast Asia was discovered in 1984 beside the Mekong in Thailand. *The green streaks around the entrance to the old shafts are copper ore, or malachite.*

Among the most intriguing bronze artefacts in Southeast Asia are the drums found on the island of Alor, in eastern Indonesia. Their origin is unknown. *The motifs cast on to the drums suggest that they may be derived from the famous Dong Son drums which appeared in many parts of Southeast Asia more than 2000 years ago.*

must have been expensive to acquire and prestigious to own, as they seem to have remained in the possession of some groups for long periods. What was being traded for these drums is not clear, but it may even have included people. Such trade reflects the dramatic growth of centres of power, and the expansion of contact between them, that followed domestication and the introduction of metals into daily life.

Even today, bronze drums are revered and used in ceremonies in some places in Southeast Asia. One is the small island of Alor, north of Timor in eastern Indonesia. Here drums have been kept in families for hundreds of years. In the past, a man could not marry unless he had acquired at least one drum, either by inheritance, or warfare, or perhaps from the family of the bride-to-be. Today the drums are brought out occasionally as demonstrations of a family's prestige, or as the centrepiece of important ceremonies. Then one or two drums will be carried in procession, with drumming and chanting, and finally placed in the centre of the village square, guarded by priests or priestesses, while elaborately costumed dancers slowly circle round them.

One tantalising mystery about the drums of Alor is that there are so many of them on the island—certainly scores, and perhaps a hundred or more—and yet there is nothing like them on Timor or any other island nearby. All the Alor drums are similar in general shape and decoration, and the motifs on them suggest that they may have been derived from the Dong Son tradition. There is no suggestion of any bronze-making tradition on Alor itself, but they might have been made in Java, where bronze-casting was practised. One explanation for their presence on Alor is that they were salvaged by the islanders, some time in the past, from a passing trading ship that was wrecked on Alor's steep coast in a typhoon.

The drums of Alor represent virtually the eastern limit of the Southeast Asian bronze age—but not quite. In the 1950s one isolated tribe in Irian Jaya (western New Guinea) was found to possess three eroded but typical bronze Dong Son drum heads. These were presumably traded in from the west at some stage. Elsewhere in New Guinea fragments of bronze have occasionally turned up, usually carried by tribesmen as talismans, wrapped in cloth. But the most surprising discovery so far in this area was made in 1985 by Wal Ambrose, an archaeologist from the Australian National University.

Ambrose has done much of the work of identifying the sources of obsidian circulated through the Melanesian trading networks. While excavating on Lou Island, north of New Guinea, he found a small piece of bronze in a context which meant that it had been dropped there at least 2000 years ago. The nearest source of metal of this age is Indonesia or the Philippines, many thousands of kilometres away to the west. Ambrose's find is the most easterly occurrence yet recorded of products from the Southeast Asian bronze age. It is also a confirmation of the enormous extent of the trade and reciprocal flow of goods and people that grew from the domestication of plants and animals, and the dawn of the age of metals in the western Pacific.

I f our understanding of the coming of metals to Southeast Asia has
been hampered by the paucity of archaeological and physical
evidence, the same cannot be said for the other great centre of
technological innovation in the western Pacific—China. Here there
was, had we known it, a remarkable record of achievement in metallurgy
going back at least 5000 years. But it had remained largely hidden behind
China's hundreds of years of self-imposed isolation, and more recently
ignored by the West because of the long-standing diffusionist view that
China had inherited the knowledge of bronze-making from the Middle
East, and that all its metallurgy was probably equally derivative.

In the past twenty years a stream of announcements emanating from
China, together with some visits and serious analysis by Western
archaeometallurgists, have completely transformed that picture. Today it
can be said with absolute assurance that China not only achieved its own
independent transition into the age of metals, at a time very similar to such
developments elsewhere, but in some clearly identified areas was nearly
2000 years ahead of the West.

A major contribution to this new understanding has been the painstaking
work recorded in two monumental publications.[6] One is the great series of
volumes entitled *Science and Civilisation in China*, which is being compiled
and largely written by Sir Joseph Needham, a Chinese-speaking historian
of science at Cambridge University. Needham and his team began their
task in 1948, and have so far published ten out of a projected twenty parts.

On Alor the bronze drums are
brought out for special
occasions, and sometimes form
the focus of dances and other
ceremonies. *Drums are
carefully handed down by
families, and in the past, a
young man intending to marry
had to obtain one, often at great
price, before the wedding could
take place.*

This huge bronze cauldron, cast by Shang metalworkers in China more than 3000 years ago, weighs 875 kilograms, and is the largest known metal object of that age. *It was found in 1939 by peasants working in the fields near the site of the ancient Shang capital.*

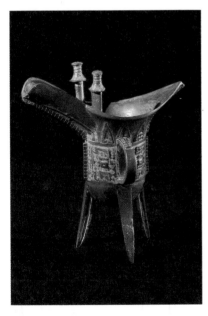

A more delicate example of the skill of the Shang metal casters is this wine cup. *The Shang bronzes were made as ceremonial objects, to be included in the burials of nobles.*

The other important work is *Metallurgical Remains of Ancient China*, by Noel Barnard, a historian at the Australian National University, and Professor Sato Tamotsu, a Chinese-language professor in Tokyo. After a lifetime spent in Chinese studies from original sources, Barnard is a leading Western authority on Chinese metallurgy. One of the most fascinating sections of his book deals with the question of whether bronze-making was independently invented in China or whether the knowledge came in from the west.

Barnard publishes maps showing all known ancient metallurgical sites in China, and matching maps showing known deposits of copper, tin and lead. The development of metallurgy shows a pronounced centrifugal pattern, with the earliest sites clustered around a core area in the basin of the Huanghe (Yellow River). This pattern matches and even overlaps the distribution of mineral resources. The further from the centre—even in a westerly direction—the fewer the mineral deposits, and the later the dates of archaeological finds. This is a picture of the way an independent Chinese technology might arise and spread, but the opposite of what might be expected if knowledge of metallurgy had come into China from outside.

Recent Chinese announcements of finds of copper artefacts and evidence of copper smelting have also gone some way towards disposing of the argument that China's bronze age could not have been indigenous, because there had been no copper age leading up to it. Finally, the publication of more information about early Chinese pottery kilns has confirmed the views of those who, like Barnard, believe that the origins of Chinese metallurgy can be found in the advanced pottery industry of the immediate premetal period, more than 5000 years ago.

It has always been obvious that Chinese potters had more efficient kilns and greater control of temperatures than anyone else. This explains why it was left to the Chinese to invent porcelain, which is made from a white clay found all over the world, but which requires a temperature of at least 1300 degrees Celsius to produce its brilliant, hard, ringing translucency. Such temperatures were unattainable in the West until a few hundred years ago. Tests have shown that pottery kilns found intact in the Neolithic village of Ban Po, near Xian, and dated to more than 7000 years ago, could maintain temperatures as high as 1400 degrees Celsius, which is more than enough to smelt copper, even without forced draught.

Further support for this theory of origin is provided by one of the triumphs of Chinese metallurgy—the bronzes produced during the Shang dynasty, which lasted from 3500 to about 3000 years ago. There is a direct continuity of form from pottery vessels of 5000 years ago to many of the characteristic Shang bronze cauldrons, bowls, and three-legged wine cups. Even more convincing is the similarity in techniques of decoration and casting. The early pottery makers used clay models, with stamped and incised patterns on them, to make baked clay moulds, from which the finished clay objects could be turned out. Precisely the same methods were used, 2000 years later, by the Shang craftsmen to cast their stunning bronzes.

Today the fields around Anyang, site of the ancient Shang capital, are an intensively cultivated patchwork of crops, growing in the rich alluvial soil of the flood plain of the Huanghe. Here the Shang ruler Pa'anken established the first recognisably Chinese civilisation. The people must have lived very much as the people of Anyang do today, by farming, but some among them found time to create one of the most powerful and technically accomplished bodies of art ever known.

The Shang craftsmen used clay moulds and molten bronze to produce, not weapons or utilitarian tools, but ceremonial vessels for use in rituals. These were primarily to do with the life and death of prominent people, and the bronzes were then buried in graves with the remains of the nobles. There were about thirty main styles of Shang vessels, ranging in size from a slender wine cup to the huge rectangular cauldron on four massive legs that was unearthed by peasants near Anyang in 1939. It weighs 875 kilograms, and is the largest metal casting anywhere in the world which is more than 3000 years old.

One intriguing aspect of the Shang bronze story is that these remarkable works of art became well known to museums and collectors around the world a long time before anyone knew anything about the culture which created them. This was because, from the earliest times, looters had been digging up the graves of the Shang nobles and recovering the bronzes. Many were sold as scrap metal and melted down, while others were taken out of China by explorers, missionaries, and foreign soldiers of fortune.

But beginning in the late 1920s and continuing until the 1950s (interrupted only by the war with Japan), systematic excavations by Chinese archaeologists uncovered not only the ruins of the ancient Shang capital, but even more striking examples of bronze craftsmanship than those upon which its fame had rested. Today the collections of Shang bronzes in China's own museums can match the best examples anywhere from this amazingly creative period in its history.

After the downfall of the Shang dynasty in 1027 BC the Chinese bronze-casting tradition continued for another 1000 years or so, but with a steady decline in artistic expression and technical casting skills. It was, however, based on a highly organised copper and tin mining industry, whose scale and capacity is only now becoming clear. In 1987 we were able to film at a site of ancient copper mining which had only just been opened to foreign visitors, but which has been attracting archaeometallurgists from all over the world ever since.

The mine is at a place called Tonglushan, near the Yangtze River in Hubei Province. The name in Chinese means 'green mountain' because, a long time ago, it was noticed that after heavy rains the mountain showed patches of green, where outcrops of malachite (copper oxide) had been washed clean. Today the mountain has gone, and in its place there is a huge open cut copper mine, which is producing vast amounts of copper. But in recent years, as the huge power-shovels ate away at the mountain, unusual shafts and tunnels sometimes appeared in the walls of the open cut. Archaeologists from the Provincial Museum were called in, and found an extraordinary chronology of copper mining and smelting, going back 3000 years.

So far the investigations have discovered several hundred vertical shafts, nearly 100 drives or horizontal tunnels, twenty-five smelting furnaces, and ancient smelting slag estimated to total 400 000 tonnes. There are mining works, structures and tools from many different dynasties: furnaces from the Sung dynasty 1000 years ago, mining tools from the Han period, 2000 years ago, and, most remarkable of all, timber-lined shafts and drives from mining operations more than 3000 years ago. These are the world's oldest mining structures to survive in anything like their original form.

The early Tonglushan workings were preserved by mud, which filled the shafts and drives. When cleaned out, they yielded an amazing variety of mining tools: bronze adzes, chisels and pickaxes used by the miners at the ore face, and wooden shovels, rakes and sleds for dragging the copper ore to the shafts for hauling to the surface in baskets of wood and leather. Perhaps the most interesting surviving artefacts are the timber roof beams and supporting posts, which in many drives are preserved intact. They are joined by neatly cut mortice and tenon joints.

A spectacular area of the exposed workings has been enclosed inside a huge hangar-like museum in the middle of the present mining operations. In the nearby town of Daye there is a full-scale reconstruction of the

A large section of the ancient mine workings at Tonglushan, in China, exposed by modern operations, has been excavated and enclosed in a museum on the mine site. *These workings are more than 3000 years old, but the wooden posts and beams, with their mortice and tenon joints, are remarkably well preserved.*

Each of this extraordinary set of Chinese bronze bells, 2400 years old, plays not one but two distinct notes, depending upon where it is struck. *The whole set of sixty-four bells, discovered in a 2500-year-old tomb, covers a range of five octaves, and was used as an orchestra.*

underground shafts and drives, showing miners at work, digging ore, testing samples and hauling material to the surface. It shows a pattern of working that continued, with very little change, into modern times. In many parts of the world men are still digging out mineral wealth from shafts and drives just like these. Together, the two museums at Tonglushan demonstrate a scale and development of copper mining in China 3000 years ago that was not matched anywhere else in the ancient world.[7]

The Hubei Provincial Museum at Wuhan displays a very different but no less astonishing side of Chinese metal technology, which again has no counterpart in the West.

In 1978 archaeologists working near the Yangtze River, on the oppposite shore to Tonglushan, discovered the tomb of Yi, the ruler of one of the Warring States, who was buried 2400 years ago. The site was waterlogged, and this condition had preserved in the tomb an amazing collection of wooden articles, including musical instruments and lacquerware. But the most startling find was of a large wooden frame, with elaborately cast bronze ends, to which were attached sixty-four large bronze bells. The bells were graduated in size, and ranged in weight from 2.4 kilograms up to a massive 203.6 kilograms. They were beautifully cast, with elaborate decoration and inscriptions, but were unusual in shape, being oval in section, with the rim cut away in a shallow arch on the two long sides. Like most Chinese bells, they had no clapper but were obviously meant to be struck with a mallet or pole.

Single bells of a similar shape had been found in China before, but never a complete set like these, and the staff at the museum in Wuhan were excited by their acquisition. The gradation in size, the way the bells were hung, and the ancient inscriptions on the bells themselves suggested that they were meant to be played like a set of chimes. It was only when the museum personnel began to play the bells, however, that they discovered just how unusual these apparently simple castings were.

Each bell from the monarch Yi's tomb, depending upon where it is struck—at the centre of one side, near the rim, or higher up, near the shoulder—plays not one but two distinct notes. Other bells, no matter what their size or shape, and regardless of where they are struck, produce only a single note. The interval between the two notes is either a major or minor third, equivalent to a difference of four or five notes on the piano. The reason for this tuning system was apparently to enable the whole set of bells, covering five octaves, to be played as an orchestra for their ruler.

Chinese archaeologists, acoustic scientists and foundry engineers spent five years casting exact replicas of the original bells, which are now on display in a sealed room in the Wuhan museum. After exhaustive analysis and study of the replicas, they have worked out the extremely sophisticated acoustic design which produces two notes from a single casting. The two critical factors are the oval shape and the arched rim. Working together, they produce two completely different sets of harmonic vibrations, depending upon which part of the bell is struck. The single casting thus behaves as if it is two separate bells.

The second and equally masterly technical feat of the ancient bell-makers was in achieving perfect tune for the two notes they selected for each bell. Tests with the latest electronic equipment show that the frequencies are as close to the correct pitch for each tone as a modern piano. Today, bells are finally tuned by shaving metal from the inside, after casting. The bells from this tomb show no signs of any such tuning. They must therefore have been designed and cast from moulds with such precise tolerances that the frequency response of the finished casting is perfect—not just for one note, but for two. The whole concept and execution involved a knowledge of theoretical acoustic principles and practical metallurgical quality control, 2400 years ago, that now seems almost impossible to credit. And the secrets of that knowledge were lost and forgotten, soon after the dead king was buried, not to be resuscitated until the discovery of the tomb in 1978.[8]

The final dimension of this astonishing achievement of Warring States technology lies in the performance of the bells. The staff of the Wuhan museum now give regular recitals of both ancient and contemporary music. They use up to three players with pairs of mallets to play in the middle and upper ranges of the five-octave set, and one with a heavy pole to play the great bass bells. The sounds are clear and mellifluous, and effortlessly bridge the interval to a far-off, shadowy period of China's history, transporting the listener to the court of Yi in a way that words or even paintings could not hope to do.

This and other brief references we have made to Chinese bronze age accomplishments give only a hint of the output of the inventive Chinese mind over the past few thousand years, which helped to make that vast country for so long the powerhouse of the Pacific. In our next chapter we will look at some of the more practical expressions of that energy and imagination, including China's unmatched innovations in the use of the universal metal, iron.

9.

The Powerhouse

For more than half a million years, since those early people sat around their fires in the caves of Peking Man at Dragon Bone Hill, China has been at the centre of human development on the western shores of the Pacific Ocean. This vast expanse of mountains, rivers and fertile plains, extending from the tropics to the frigid borders of Siberia, was one of the first places in the world to provide a settled life for hunter gatherers, and to witness the growth of large human populations. It was from this Asian heartland, between 50 000 and 100 000 years ago, that the first adventurous groups of modern people began to move out to occupy the vacant territories of the Pacific basin. And in more recent times, over the many thousands of years of recorded history, there has been no greater concentration of human energy, intellectual and physical, than that contained in China—the powerhouse of the Pacific.

China has huge material resources, but its domination of the Asian region is deeply rooted in its human resources. Very early the Chinese people developed a remarkable capacity for technology, which they adapted with great imagination to suit their philosophy and the way they chose to arrange their society. That combination raised the Chinese to leadership in many areas of human activity—social, cultural and scientific—and not just in Asia, but in a worldwide context. But that superiority was not maintained. A few hundred years ago, for reasons which are still obscure, this great nation slowly began to run out of energy, and went into decline. By the time the Industrial Revolution swept the West, China's technological innovations had been taken up everywhere, and their origins largely forgotten. The most populous nation on earth was eclipsed. Only now is it emerging once again as a world power, regenerating the dynamism which maintained the Middle Kingdom for so long at the centre of the universe.

There are many ways to look at the Chinese and what made them different from everyone else. Some of those differences are cultural and do not translate easily. But there is one particular facet of the Chinese

The revolution in agriculture in China in the last few centuries before the Christian era triggered the surge in population growth which made that vast country the powerhouse of the Pacific. *Workers bring in the rice harvest in Sichuan, the 'ricebowl of China'.*

character that expresses their dynamism in a tangible and practical way, and that is their extraordinary record of invention and innovation. So pervasive in modern life today are the discoveries made in China that it is sometimes hard to believe where they came from. It is impossible here to give even a basic list of those inventions, but we would like to describe some which had particular significance for China itself.[1]

China's record of innovation goes back to Neolithic times, when it was the scene of some of humanity's earliest experiments in the domestication of plants and animals. Since then the Chinese have devoted a great deal of attention to the relationship of people to the land, and to their crops and animals. (China is still an overwhelmingly rural society, with eighty per cent of the population classified as rural in the 1982 census.) Yet the Chinese have also an aesthetic side to their very practical nature. One of their oldest and most famous rural inventions was first a curiosity, then a luxury, and only later an industry with worldwide ramifications.

The story begins about 5000 years ago when, according to Chinese legend, the wife of the mythical Yellow Emperor, Huang-ti, was resting one day with her ladies-in-waiting under a mulberry tree in the palace garden. On the leaves of the tree she noticed a number of fat white caterpillars, weaving their heads back and forth. Observing them closely, the empress saw that they were drawing fine, lustrous threads from within their bodies and spinning cocoons for themselves. The empress picked one of the finished cocoons off the tree and tried to pull the threads free, but they were firmly stuck in place. While she was struggling with the cocoon it dropped into her cup, which had just been filled with hot tea. As she tried to fish it out with a twig she caught the end of a fine thread, which had floated free as the tea dissolved the glue holding it. As the empress cautiously pulled at the thread, it unwound from the bobbing cocoon. She then made the astonishing discovery that each cocoon was wound from a single, continuous thread. Soon the empress and her ladies-in-waiting were all soaking cocoons in hot tea and winding the threads on to twigs. Eventually, so the story goes, they collected enough threads to weave a beautiful robe for the emperor—and for centuries afterwards the making of this special cloth was confined to the royal household.

Whether or not this is how silk was first woven into cloth, China was certainly the place where it happened. And China is still the world's largest silk producer. Each spring a green carpet unrolls across the warm, humid coastal plains of eastern and southern China, as millions of mulberry trees burst into leaf. Both the white mulberry *(Morus alba)* and the silk moth *(Bombyx mori)* originally lived wild in China, and at some stage the moth developed a particular relationship with the mulberry. Although the caterpillar stage of the moth, the silkworm, can survive on a number of different plants, it only produces silk when it eats the leaves of the white mulberry. For this reason both trees and silk moths are now intensively cultivated in China, Japan, India and other silk-growing countries.

In China, in what is essentially still a cottage industry, peasants raise their silkworms at home and grow plantations of mulberry trees to feed

The production of silk from the cocoons of the silkworm moth was invented in China, and China remains the world's largest producer. *Farmers in Chekiang Province take their cocoons by boat to the central collecting stations.*

them. They keep their growing silkworms on large trays made of plaited grass, and collect fresh mulberry leaves for them every day. In a closed room the munching of thousands of tiny jaws sounds like the faint but incessant crackle of radio static. In their growing stage the caterpillars eat nonstop, consuming something like fifty times their own weight of leaves. They are then provided with small tents of straw to climb on and spin their cocoons. About five days later the harvest begins. The first harvest, in spring, produces the best silk, but there is another harvest in midsummer and two more in autumn.

Silk production in China still depends almost entirely on individual farming families and at harvest times they bring their cocoons to central collecting stations run by the local government or their own cooperatives. All across the landscape, along lanes and canals between the mulberry groves, the huge baskets of creamy white cocoons can be seen converging on the collecting stations, swinging from shoulder poles, balanced on the seats of bicycles, stacked on the decks of barges.

At the collecting stations the cocoons are dumped into large oval baskets and weighed, and the owner credited with the seasonal price. After many centuries of selective breeding the cocoons are all of uniform size, shape and colour, but a small sample of each load is tested for quality, to cater for small variations within the permitted range. The bulk of the cocoons brought in go straight to the silk factory nearby. Some, however, are put into store to keep the spinning machines running during the winter. To prevent the pupating moths from hatching they are quickly killed by steam heating.

The silk factories across China, whether large or small, are all laid out on the same pattern, with long rows of winding machines, each tended by a nimble-fingered operator. The first process is the unwinding of silk from the cocoons, and this is still done essentially the way that the Yellow Empress and her handmaidens did it. The cocoons are tipped into a bowl of hot water beside each operator, twenty or so at a time, to melt the glue

In the silk mills the cocoons are placed in hot water to soften the glue which binds the silk, so that the single, continuous thread can be wound off on to spools. *The thread on each cocoon is about one kilometre long.*

which binds the thread. The ends of the threads float free and are picked up by mechanical fingers on the end of a shaft, rotating in the bowl.

Dextrously the operator catches the ends of the threads, then empties the cocoons into another trough of hot water in front of her, full of bobbing cocoons that are already being wound off on to spools above her head. Her task now is to add the threads from the fresh cocoons to those being wound on to the spools. But although a thread of silk is stronger than any other natural fibre, or even iron wire of the same thickness, it is finer than a human hair and liable to be broken by the pull of the spindle. So the operator takes six or eight filaments and twists them together, before deftly joining them to one of the high-speed spools. Each cocoon produces a single, unbroken thread up to a kilometre long, and when that is wound off the pupa is discarded. In some parts of China the pupae are deep-fried and eaten as a delicacy; in others they are pressed to extract a rich, smooth oil that is used cosmetically.

In each huge room of a silk factory there are hundreds of unwinding tanks and spools operating simultaneously, with steam rising from the tanks where thousands of cocoons are bobbing and spinning. With the incessant clatter of the machines, the creamy gleam of the silk building up on the spinning spools, and the rich smell of warm cocoons, the atmosphere is overpowering.

Although all the silk grown in China is a very pale cream, almost white, the Chinese discovered very early in their use of this natural filament that it soaks up colours, evenly and permanently. This, together with its natural sheen and elasticity, made silk a weaver's dream. And so, several thousand years ago, there came into existence in China the fabric which made China a legend in the West, long before anyone really knew where this mysterious country lay. It is not certain when silk cloth, with its luxurious sheen and brilliant colours, first began to reach the West, but when it did the demand for it in the great cities and empires around the Mediterranean could never be satisfied.

Many traders were prepared to face the dangerous 12 000-kilometre journey across mountains and deserts to China, to offer any price for lengths of silk. The route they took, which came to be known as the Silk Road, was the only link between the great empires of China in the East and Rome in the West. But although silk worth huge amounts of gold travelled that route, the knowledge of how it was made did not. The Chinese would sell silk, but would never disclose how it was made. There was great speculation in the West, but complete bafflement. Aristotle perhaps came as close as anyone, when in the third century BC he suggested, in his book *The History of Animals*, that silk was made by some kind of insect.

The Chinese guarded the secret of silk for thousands of years—at least until after the beginning of the Christian era. The penalty for trying to take silkworms or mulberry trees out of China was death. It was not until the second century AD, according to Chinese legend, that the secret was taken to Korea as part of the Buddhist religion. From Korea both mulberry trees and silkworms reached Japan, and were later smuggled into India. But the knowledge did not reach the West until the sixth century, in a clandestine manner.

At that time the Emperor Justinian and the Empress Theodora were ruling what was left of the Holy Roman Empire from Constantinople (now Istanbul). Constantinople, straddling the Bosporus and linking Asia and Europe, had been a great distribution centre for silk from the East, and Justinian had just taken control of the trade for himself when a long war with Persia cut off all supplies. So the emperor was most interested when he was approached by two monks who had recently returned from China. When they told him they had found out how silk was made, Justinian offered them a huge reward to go back and smuggle silkworm eggs and mulberry seeds out of China. They did so, inside their hollow walking-sticks, and for a time Justinian and Theodora had a monopoly of silk production in the West. Today, of course, silk is produced in many countries, but no one has yet been able to emulate the mystique or reputation of

The first woven silk was made in China more than 5000 years ago. *These fragments, now in the Beijing Historical Museum, were found in a tomb from the Han dynasty (206 BC to AD 220).*

Chinese silk. In Chinese silk mills we saw strictly guarded corners where special orders are woven or printed for the great couturiers of Europe and the United States.

The Chinese still maintain close links with the origins of their great discovery, through the work of the Silk Research Institute in Nanjing. Here the staff and students maintain the ancient traditions of silk weaving, using wooden machines copied exactly from tomb drawings of early looms—some from the Han dynasty, more than 2000 years ago. They learn the basic skills of those first weavers, then adapt their ideas to new patterns. They also reproduce the intricate designs that went out from China all those thousands of years ago along the Silk Road.

Despite the size and complexity of the Chinese silk industry today, the past is never very far away. In the silk-producing cities of Hangzhou and Suzhou we saw great barges loaded down with bales of silk setting out along the Grand Canal on the first leg of their journey to Shanghai and then on to foreign markets—just as they had done 2000 years before, on their way to the starting point of the Silk Road in Xian.

It was an appropriate combination, the Silk Road and the Grand Canal, because while the former was China's first link with the outside world, the latter was the first transport link of any real significance inside China itself. The Grand Canal was a unifying channel for goods and ideas that was crucial to the growth of the Chinese state. Besides the economic benefits that it brought, successive Chinese rulers realised its social significance as a conduit for the exchange of news and information and the strengthening of the concept of national unity. It is in fact almost impossible to overstate the importance of the role of the Grand Canal in China's political, social and economic development.

The Grand Canal (Da Yunhe) meanders from north to south through a huge and very important part of eastern China, crossing and linking the Huanghe (Yellow River), the Yangtze, and three other major rivers. It connects Beijing in the north with Hangzhou, 1800 kilometres to the south. Altogether, including branch canals, there are more than 2700 kilometres of waterways, serving some of China's most important industrial cities in four provinces. Many civilisations have dug canals, for irrigation or transport, but none on a scale anywhere near as large as this.

The Grand Canal was begun in the south 2400 years ago, when the ruler of the state of Wu, Fuchai, ordered the digging of a canal to link the

Bales of raw silk being loaded on a barge for shipment along the Grand Canal, just as they were 2000 years ago. Then, the silk completed its journey to the West along the Silk Road. *Although silk is now produced in many countries, China remains the major exporter.*

Yangtze and the Huaihe River, as a way of extending his influence. From that time on the canal steadily grew, through many different dynasties. In the early seventh century, during the Sui dynasty, the decision was taken to extend the system to provide an inland water-transport route from the great food-growing region south of Shanghai to the north, where the severe climate caused frequent food shortages. Some five million people were conscripted to work on the project. As many as two million may have died in the course of the work, but eventually the canal was cut in a huge V-shape, from Hangzhou northwest to the Sui capital of Luoyang on the middle reaches of the Huanghe, and then northeast to Zhuojon (now Beijing). When the Yuan rulers moved the capital to Beijing in the late thirteenth century they decided to straighten the Grand Canal, in order to shorten its length of 2500 kilometres. Kublai Khan, the first Yuan emperor, spent eleven years supervising the cutting of the present canal, which runs directly south from Beijing to Hangzhou. At one stage he ordered all the senior government officers, including the Prime Minister, to join the workforce.

With the decline of the Ching dynasty during the nineteenth century, and the wars and turmoil of this century, sections of the canal fell into disuse. Parts of it were destroyed by catastrophic floods on the Huanghe (known, for this reason, as China's 'river of sorrow') but large stretches remained intact, and today it is being brought back into use with a massive expenditure of money and labour. When fully restored, the Grand Canal will be open again to shipping from Beijing to Hangzhou, regaining its position as one of the great engineering achievements of the ancient world.

The Chinese interest in water transport produced many innovations, which were to be seen along the Grand Canal but appear to have been unknown outside China at the time. One, invented in the fifth century, was the paddlewheeler. By the twelfth century these had grown into huge warships driven by a large stern paddlewheel. Some were nearly 100 metres long and carried 800 men on as many as ten decks. They were propelled by up to 200 men, working treadles.[2]

Another key Chinese invention linked to the Grand Canal was to have worldwide application, but not until much later. This was was the pound-lock, or water-lift, which solved the problem of joining waterways at different levels, and of building canals across undulating country. It was in AD 984 that Chiao Weiyo, an assistant commissioner of transport during the Sung dynasty, conceived the brilliantly simple idea of using two pairs of gates or doors to form a step between a higher and lower waterway. When a vessel on the lower waterway wanted to pass up the lock, the upper gates were closed and the lower ones opened. The vessel entered the lock, the lower gates were closed behind it, and water was let in from the upper level. As the water level rose the vessel rose with it, until it could sail out of the lock at the upper level and continue its journey. The reverse passage could be made just as easily, by draining water out of the lock with the vessel in it.

Li Bing was the governor of Sichuan Province when, in the third century BC, he conceived and built one of the world's greatest irrigation projects. *This sandstone statue of Li Bing was found in the bed of the Minjiang in 1974.*

Chiao Weiyo's most elegant refinement was to make the pairs of gates meet at an angle, pointing upstream, so that the pressure of water did all the work of keeping them tightly closed. The Grand Canal system ultimately depended upon locks, hundreds of them—more than sixty in a single 100-kilometre stretch. Curiously, by the time the lock was adopted in Europe in the fourteenth century its value on the Grand Canal was already declining, as sea transport was beginning to take over the work of moving grain and other goods from southern China to the north.

In keeping with the wall of mystery which has surrounded China for much of the past 5000 years, and the consequent Western ignorance of so much of Chinese innovation, it seems appropriate that one of the most astonishing and visionary feats of technology ever carried out in that country should still be virtually unknown outside China. This is all the more remarkable since the Dujiangyan Project is still working as well as when it was constructed more than 2300 years ago.

The scene of this remarkable project is Sichuan Province, in the southwest of China. Sichuan is a very mountainous region, but its heart is a vast plain set in a bowl of hills. This plain, around the provincial capital Chengdu, is one continuous carpet of irrigated farms, growing rice, wheat, oilseeds, fruit and vegetables. It is the richest food-producing zone in China, and Sichuan, with 100 million people, is the most populous province and has the highest per capita income. And the roots of this prosperity can be traced to the farsightedness of the governor of the region in the third century BC, an engineer named Li Bing.

Although China had yet to be unified, it was a time of rapid economic development. The basis of this was a drive in many of the individual kingdoms to increase food production by clearing land, draining marshes and areas of salinity, and introducing improved farming practices such as soil classification, the use of manure, and optimum timing of ploughing and sowing. The key to it all was a series of great irrigation projects—not so much to ameliorate the effects of drought, although that was important, but to bring more land into production. And the greatest of them all was the Dujiangyan Project, conceived by Li Bing.

The Minjiang (Min River), emerging from the mountains northwest of Chengdu, provides the major drainage in the Sichuan basin, but historically its contribution had been chiefly in the form of catastrophic floods. Li Bing conceived the idea of diverting water from the Minjiang to irrigate the fertile but dry plains. In an age before earth-moving machinery, reinforced concrete or steel pilings it was a daunting challenge to harness a river that was several hundred metres wide and poured out of the mountains, when in flood, like a mill-race. To maintain the regular, year-round flow on which irrigation depends, Li Bing had to find a way of taking most of the Minjiang's flow in the dry season, but only part of the flood when the river was high. Amazingly, he got the whole thing right, first time.

Li Bing selected a place called Dujiangyan, where the Minjiang was separated from the flat plains only by a rocky ridge called Jade Rampart Mountain, and assembled a huge workforce of tens of thousands. He set

The Grand Canal, begun 2400 years ago, had a profound influence on the unification and economic development of China. *This section of the Grand Canal is in Songling, a small silk-producing centre in Jiangsu Province.*

them to making thousands of large bamboo baskets and filling them with granite boulders from the river bed. Then, during the dry season, they began dumping the heavy baskets on a long sandbank that ran down the centre of the river. By the time the floods came and covered their work, they had laid down a foundation of rock that did not wash away. After several dry seasons, Li Bing had created an artificial island, more than 100 metres wide and a kilometre long, that split the Minjiang into two channels.

The western channel became the main course of the river but was controlled by sluice gates, so that any desired proportion of the flow could be diverted into the eastern channel. This carried the water that Li Bing planned to use. Next, he had his workers cut a passage through the solid rock of Jade Rampart Mountain, to let the water from the eastern channel of the Minjiang flow out on to the plains. He kept the flow constant through this gap, known as 'the mouth of the precious jar', by diverting more or less water from the western channel, depending on the season.

The water from Li Bing's project brought more than eight million units of new land under cultivation. There were so many irrigation channels that the region became known as the 'sea on land'. Within a few years the food

Li Bing's great irrigation project, built 2300 years ago, is still working perfectly, irrigating a huge area of Sichuan Province. *These channels carry the water through the city of Dujiangyan and out into the farmlands.*

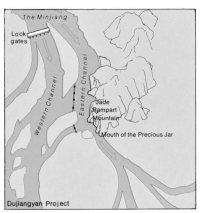

that it provided enabled Li Bing's ruler in the kingdom of Chin to raise and maintain a large army and defeat all his rivals. In 221 BC the country was unified under its first emperor, Chin Shih Huang, whose kingdom was to give its name to a nation and a people.

Today steel and concrete have replaced the bamboo and stone walls on Li Bing's island, but the vast project is still working, after 2300 years, just as he conceived and built it—one of the world's most successful and enduring works of hydrologic engineering. In 1974, during maintenance work on the western channel, workmen found a sandstone statue of Li Bing, face down in the bed of the river, where it had lain for 1800 years. It stands today in a temple beside Jade Rampart Mountain, a lifelike figure of a calm, smiling, confident man, one of China's most influential but least-known founding fathers.

A city has grown up at Dujiangyan, and when the summer rains swell the Minjiang the torrent pouring through the 'mouth of the precious jar' thunders past tall buildings and under crowded bridges on its way out to the fields around Chengdu.

We followed the water, as the channels were divided and subdivided, until it finally trickled around the roots of rice plants, heavy with grain. Today Li Bing's original irrigation area has been expanded to include more than three million hectares of highly productive land. It enables the farmers here to harvest three crops a year from the same soil—rice in summer, wheat in winter and oilseed in spring. As we walked around the farms we saw many other inventions, most of them quite simple, which added to the success of China's agricultural growth 2000 years ago, and which are still in use because of their efficiency.

The orderly pattern of Chinese farms, with straight rows of grain and neatly laid-out plots of vegetables, was introduced in the sixth century BC (although such an approach to farming was not adopted in Europe until the eighteenth century AD). With cheap, mass-produced tools such as hoes and spades that were becoming available (of which more later), this layout

The water-lifter, made entirely from scrap wood, is a simple but efficient Chinese invention for raising water from the irrigation ditches on to the fields. *This couple is using the lifter on their farm in Sichuan Province.*

Among China's many agricultural inventions was the wheelbarrow, first used there 1500 years ago. *This device was developed from a two-wheeled cart, for use on the narrow paths between the rice paddies.*

enabled the farmers to plant seedlings and to weed the beds more efficiently. The multiple seed drill had just been invented in China, and this, too, added to efficiency—especially when the drill machine was pulled by a horse with the newly invented horse collar and chest harness. For negotiating the narrow, raised paths between the fields the Chinese progressively modified their carts, until someone finally eliminated all but one central wheel—and invented the wheelbarrow. Again, it would be another thousand years before these devices would be seen in the West.

In places we watched farmers using wooden chain pumps to lift water from the irrigation channels into their slightly elevated fields. This pedal-powered device, invented 2000 years ago, is made entirely of scrap wood. Square wooden plates attached to an endless chain made of wooden links move through a wooden tube, which has one end in the water and the other over the edge of the field. Although the plates fit the tube quite loosely, they move a constant flow of water up the tube and on to the field. The effort needed to drive the chain by two sets of pedals is well within the power of the farmer's wife and one of his children. (This device may have given rise, in the tenth century AD, to another Chinese invention, the steel link chain. This was used in China for driving clockwork but was taken up in the West for its best-known application, the bicycle.)

We saw other machines still in use after 2000 years—an efficient foot-powered rotary thresher which stripped the sheaves of rice in seconds; a hand-driven rotary fan for winnowing the rice grains from the chaff; and a stone husking mill driven by a horizontal waterwheel. The great stone wheel, mounted at the end of a short arm, trundles round in a circular trough, into which the rice is shovelled. Within a minute or two it grinds the husks from the kernels. Because the arm is mounted directly on the shaft of the waterwheel below the mill floor, there is no need for complicated transverse gears, such as Western watermills require.

Virtually all the agricultural machinery we saw was powered by water, animals or human muscle, which has always been in good supply in China. It was the combination of all these factors—and the work of imaginative engineers like Li Bing—which set in motion around the third century BC a development of the utmost significance for the whole of Asia and the western Pacific. That was an irresistible surge in the growth of China's population. When the world's first census was held during the Han dynasty, in the second year of the Christian era, China already had a population of nearly fifty-eight million taxpayers—more people in one nation than in all the territories of the Roman Empire.

The dramatic strengthening of centralised power in China was based on the tremendous economic and social benefits of improved agriculture. And this in turn was largely due to the decisive role played by one group of Chinese technologists during the Han period. They were the iron masters, and their discoveries of how to cast iron and make steel gave China an enormous advantage. The great divergence between the technologies of East and West really began with iron, and the reasons why each went their separate way are fascinating.

A key factor in the growth of China as a major power in the Pacific basin was its huge population. *In the world's first census, held in AD 2, China already had a population of nearly fifty-eight million taxpayers.*

In the Middle East and the Mediterranean the iron age replaced the bronze age only because of a crippling shortage of bronze which developed towards the end of the second millennium BC. In the chaos which swept across the civilised world of the eastern Mediterranean around 1200 BC, with the attacks of the 'Sea People' on Greece, Anatolia, Crete, Cyprus and the great ports on the Levant, established trade routes with the ancient world to the east were cut and the supplies of tin broke down. Without tin there could be no bronze, and once the existing supplies of bronze had been recycled to their limit the bronze age ended, for ever. Metalsmiths turned to iron, which was common—but, at first, vastly inferior to bronze.

Iron was made by smelting in furnaces, more or less as copper had been smelted. But because the metalsmiths had no way of heating their furnaces to the melting point of iron—around 1540 degrees Celsius—they could never produce liquid iron. All they could get was a 'bloom'—a lump of slaggy material containing iron particles. This had to be laboriously hammered to get rid of the slag. The result was a piece of pure iron, which could be heated and shaped by hammering, but not melted or cast. Wrought iron, as it was called, would not hold an edge like bronze, and it rusted very quickly. All in all, it was an inferior material for tools and weapons. Eventually, blacksmiths learned how to improve iron in three ways: by heating it in a forge they added some carbon to it; by heating and then suddenly quenching it they made it harder, but brittle. And by reheating it somewhat less (tempering) they made it less brittle but left it still strong. In fact, they learned how to turn iron into steel. But for the next 1500 years that was all they could do with it, and everything they made with iron or steel had to be done by hammering.

The Chinese iron masters were able to take an entirely different road, because of one critical difference in technology: their more efficient furnaces, derived from their pottery tradition, and further improved by two crucial Chinese inventions—the horizontal bellows and the double-acting box bellows. The blacksmiths' bellows used everywhere except in China worked on a concertina principle, with two flat pieces of wood hinged at one end and joined by a leather membrane. The blacksmith or his helper raised and lowered the handle to squeeze air into the forge. This limited the size of the bellows—and their blowing capacity—to the weight of the beam that a man could lift up. The Chinese also made their bellows like a concertina, but in the shape of a collapsible leather barrel. They hung it on its side from a beam and squeezed it horizontally. The total weight, being suspended, was immaterial, and all the effort could be put into the horizontal push. Such bellows were made very large, and driven by groups of men, animals, or even a waterwheel and piston.

Even more unusual, and crucial in the Chinese iron age, was the double-acting box bellows. This ingenious device had a piston inside a square box, connected to a handle projecting from one end of the box. When the handle was pushed or pulled, a simple system of valves diverted air through a pipe into the furnace on both the push and pull strokes. The result was a virtually continuous flow of air to the furnace, with the minimum effort. When scaled up in size, and driven by a team of men, this bellows could operate a blast furnace several metres high, capable of handling a charge of several tonnes of iron ore and charcoal. In the presence of such large quantities of fuel the iron absorbed carbon, and its melting point was considerably reduced. The iron which flowed out of the furnace was high in carbon, brittle, but easy to cast.[3]

With this furnace technology, the Chinese could not only smelt iron in large quantities but also cast it in moulds, as they had cast bronze. And from about the third century BC they began to mass-produce iron tools. Using multiple-piece iron moulds they turned out millions of cast iron ploughshare blades, hoes, spades, wheel hubs, harness pieces and many other tools for the expanding agricultural sector.

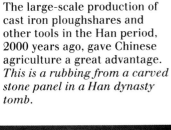

The large-scale production of cast iron ploughshares and other tools in the Han period, 2000 years ago, gave Chinese agriculture a great advantage. *This is a rubbing from a carved stone panel in a Han dynasty tomb.*

The invention of the double-acting box bellows gave Chinese metalworkers the power to make cast iron 1500 years before it could be produced in the West. *The handle is connected to a piston inside the box, and by a system of valves blows air to the furnace on both the push and pull strokes.*

The large blast furnaces developed in China during the Han period (206 BC to AD 220) enabled the iron masters to turn out vast numbers of iron tools. *A drawing from the Honan Iron Museum near Zhengzhou shows the ironworkers at the front of the furnace, tapping the molten iron into moulds in the sand. Ingots of cast iron are piled next to the moulds.*

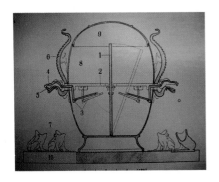

An ingenious Chinese invention was the earthquake detector. The column balanced in the centre was dislodged by an earth tremor and fell along the line of the shock wave, activating a rod which released a bronze ball into the mouth of a bronze toad, thus alerting the officials and indicating the direction of the earthquake. *This drawing is of a model in the Beijing Historical Museum.*

In the Huanghe basin, where the great deposits of iron are to be found, Chinese archaeologists are excavating impressive evidence of the Han iron industries. Near Zhengzhou we saw the Honan Iron Museum, which has been erected over the excavated remains of several huge blast furnaces. (We reproduce the museum diagram, showing how the furnaces worked.) Sometimes, when a furnace broke down, or the walls cracked, waste iron collected behind the furnace lining or in the base. Then the workers had to dismantle the furnace and remove the waste iron, called a salamander. At the museum there are salamanders, abandoned 2000 years ago, that weigh more than twenty tonnes. They are unforgettable reminders of the scale and technical mastery of those Han iron masters. Their ability to cast iron was not matched in the West until the fifteenth century, when water-powered blowing engines in Italy and England finally provided the blast that was needed.

The Han technologists invented another way of using their casting technology, which was never discovered in the West: stack casting. At the Iron and Steel University in Beijing we were shown a series of small clay moulds, designed to be stacked one on top of the other and filled with molten metal in a single pour, producing up to fifty identical cast objects. Archaeologists have found thousands of these clay moulds in Han furnace sites, and they were obviously used to turn out huge numbers of small items such as harness pieces, buckles and coins.

One of the most fascinating examples of ancient technology that we filmed in China was a demonstration of Han stack casting at the Iron and Steel University, arranged for us by the Director, Professor Tsun Ko. Using metal forms made from the original Han moulds, two staff members expertly produced about fifty sets of clay moulds for harness buckles and rings. These were stacked in piles of twenty, carefully lined up so that the pouring channel was clear right down through the stack. Each stack was coated with liquid clay to seal the cracks (leaving a hole for the pouring), and baked hard in a furnace. Molten brass was then poured into the stacks. When they were broken open there emerged a gleaming pile of brass pieces, identical to those turned out in such numbers by the Han metal-workers, 2000 years before.

There is an interesting sequel to this story. When the details of the Han stack-casting technology were published in the technical literature a few years ago, an iron foundry near Guangzhou (Canton) decided to adopt the method for making small cast iron objects. The plant was obsolete and the manager needed a more efficient method of production to avoid being closed down.

We watched the result—a 2000-year-old technology working as smoothly and as efficiently as it had ever done. The differences are in execution, not principle. The clay moulds are stamped out at a ferocious rate by operators with machines. The clay-coated stacks of moulds are baked in huge gas ovens. The moulds are filled with cast iron, poured from a central cauldron of molten metal melted in an electric furnace. But when the moulds are taken outside and broken open, the only difference from the scene as it would have been in Han times is in what comes out: gas rings, Yale lock casings and ornamental spikes for railings.

One common factor in the innovations and inventions that we have dealt with so far is that they were all brought into use by the Chinese during their great period of domestic growth and development, and were only intended for the benefit of their own people. In fact the Chinese have shown little interest in spreading the knowledge of their inventions abroad, which may explain why another category of innovation in which they took a commanding lead over the West is still virtually unknown outside China.

Our preoccupation with the later European discovery and exploration of the Pacific has tended to make us forget that China, and other Asian nations, had enjoyed an illustrious maritime tradition in the Pacific long before any Europeans even knew that the world's largest ocean existed. The Chinese had been exploring and trading over much of the western Pacific for nearly 1000 years, and it was, in the end, their maritime inventions and experiments which made it possible for the European navigators to break out of the Atlantic Ocean and actually reach the Pacific.

As we saw at the beginning of our story, there are maritime traditions in Asia which go back tens of thousands of years. It was probably from bamboo rafts, or more likely from bamboo itself, that the distinctive Chinese junks and sampans evolved. Like a piece of bamboo floating in the water, they have no keel, and they are divided by bulkheads or walls into a number of watertight compartments. This design, which makes for a very

The technique of stack casting, invented during the Han dynasty 2000 years ago, is used today by a foundry near Guangzhou (Canton) to make cast iron gas rings and lock casings. *The moulds contain a number of small, identical moulds stacked one on top of the other, all filled with metal in a single pour.*

The world's first compass was made in China 2200 years ago. A spoon carved from magnetic rock was placed on a bronze plate, representing the earth. When moved it always returned to a south-pointing position. *This is a model of the original, now in the Beijing Historical Museum.*

strong and seaworthy hull, was developed in China thousands of years ago but was not adopted in Europe until the nineteenth century. At the same time the Chinese adapted the water-lifter that we described earlier in the fields of Sichuan for use on board ship, thus making the first bilge pump. One or two of these, in conjunction with the localising effect of the watertight compartments, could keep any normal leak in a junk under control.

The squared-off, flattened stern of the junk also enabled the Chinese sailors to attach their steering oar vertically, with an inboard handle or tiller, thus transforming it into a rudder. It took the sternpost rudder another 1000 years to reach the West, in the twelfth century.

When it came to sails, the Chinese were remarkably advanced in their understanding of aerodynamics. Their sails were shaped like aerofoils, to generate the greatest lift from the wind. They were also stiffened with bamboo battens, so that they kept their optimum shape, even in light breezes or if they became holed for any reason. The sails were always set fore and aft, never square-rigged, and this layout, in combination with the sternpost rudder, enabled Chinese junks to sail into the wind 1000 years or more before Western ships were able to do this. The Chinese also used multiple masts on large ships, but staggered them so that they were not in a straight line. They did this to increase the air flow through the masts and maximise the effect of each sail, so reducing the risk of being becalmed.

Finally there was the invention which is arguably the most important advance in the whole history of human travel and exploration—the compass. This had its beginnings in Chin times 2200 years ago as a little spoon cut from magnetic rock. When placed on a smooth base it aligned itself north and south. With the addition of a marked board it became the first compass. This was in the third century BC. By the seventh century AD the stone spoon had given way to the magnetised iron needle floating in a bowl of water, and finally sitting on a sophisticated compass sheet. Some time between 850 and 1050 Chinese mariners began to use the compass at sea, and the whole Chinese maritime tradition entered a new phase.

From the eleventh to the fourteenth century Chinese ships built up an impressive record of ocean voyaging, as China's rulers, encouraged by military victories in Vietnam, showed expansionist tendencies. Towards the end of the thirteenth century the Mongols built a large fleet of junks for the invasion of Java, but in the end thought better of it. In the late fourteenth century, during the Ming dynasty, the Chinese navy entered its greatest period. It numbered several thousand ocean-going ships, including 400 large warships. The biggest of all were the so-called 'treasure ships', some of which were among the largest wooden ships ever built. They had up to nine masts, and the main deck could be 150 metres long and 60 metres wide — longer and wider than a football field. They sailed with crews of up to 1000 men.

Such dimensions are difficult to accept, especially as this was 100 years or more before Columbus set sail on the first long European voyage with his three tiny vessels, all of which would have fitted easily side by side on the deck of a single treasure ship. Modern Chinese maritime authorities were satisfied that the written documents and plans from the Ming period gave a true picture, but all doubts were put to rest in 1962. During excavations of a large dock on the banks of the Yangtze in Nanjing, one of the original Ming ship-building yards was uncovered. Lying in the mud was a piece of one of those great ships, perfectly preserved. It was the rudder post — the vertical shaft which holds the rudder blade. This rudder post was eleven metres high and fifty centimetres thick. It had a slot in it for a rudder blade six metres high — as tall as three men, and about the size of a rudder on a large modern liner.

One of the Ming admirals, Cheng Ho, made a series of remarkable voyages in fleets of huge treasure ships. (Cheng himself was an unusual man; he was a eunuch, a Moslem born in Yunnan whose father was a Hadji, having made the pilgrimage to Mecca.) Such a fleet must have been an extraordinary sight at sea under full sail, and it is not hard to picture the astonishment of the inhabitants of other countries when the great ships dropped anchor.

Between 1405 and 1411 Cheng Ho visited the Philippines, Indonesia and southern India. It is possible that some of his ships might have touched the north coast of Australia, but the evidence is sketchy. Between 1413 and 1433 Cheng Ho and his ships crossed the Indian Ocean to the Persian Gulf and the Red Sea, and finally went right down the east coast of Africa and may well have ventured round the Cape of Good Hope into the South Atlantic.

Admiral Cheng Ho's orders were to combine exploration, diplomacy and trade. Cheng responded by intervening in the succession to the throne in Java, putting up pillars proclaiming that the kingdoms of Cochin, Calicut and Ceylon were vassals of the Ming empire, and sending soldiers to join in a private war in Sumatra. But the overall result of his voyages was a huge gain in prestige for China in Southeast Asia and the Indian Ocean. The choice of a Moslem as commander-in-chief and ambassador to lands

China's maritime tradition reached its climax in the fifteenth century AD, when Admiral Cheng Ho led a fleet of great 'treasure ships' on a series of seven voyages across the Indian Ocean. *There is now a special museum in Nanjing dedicated to the exploits of Cheng Ho.*

where Islam had long been established was seen as a shrewd move by the Ming rulers.

Trade was also one of the major objectives of the voyages, and among the trade goods carried by the fleets were the exquisite porcelain bowls and dishes of the Ming potteries. These were traded over a large area, and in fact they are still turning up in shipwrecks around the Indian Ocean and the western Pacific. Some of the Ming wares reached Europe, through various overland routes, and began that peculiarly European obsession with Chinese porcelain. (Which is why, of course, we call it china.)

All this exploration was based in Nanjing, at that time the seat of the Ming emperors and the capital of China. Nanjing was a huge port on the Yangtze, linked downstream by river traffic to the Grand Canal and Shanghai. In the shipyards whole armies of shipwrights laboured to build the great ships, using timber from the forests of fifty million trees which had been planted near Nanjing in 1391, in preparation for the creation of ocean-going fleets for long-distance exploration.

But not everyone shared the expansionist vision and there was opposition to these expensive investments. In 1420 the capital was moved to Beijing, and Nanjing began a slow decline. The vast shipyards closed down, one by one. Interest in exploration fell away, and after Cheng Ho's return in 1433, from his last and most adventurous voyage, everything in Nanjing just stopped. Cheng Ho died the same year. (He is still regarded as a hero in China and there is now a museum in Nanjing dedicated to him and his voyages.)

It is one of the great ironies of Pacific history that the Chinese lost interest in voyaging just as they were making such expansive and competent explorations. With the loss of that interest they abandoned the oceans to others. Their great maritime inventions had by this time spread to the West, through the Arabs to the Europeans, and the continued exploration of the globe passed into the hands of other, more adventurous people. In China itself there began a long period of inward looking, in which the world was locked out—until it burst in on a moribund, stagnant society in the nineteenth century, with disastrous results for the Chinese people.

But the practical expression of those ingenious inventions can still be seen along China's coastline and on many of its lakes and rivers. The stately junks with their deep red sails are an eloquent statement of the illustrious Chinese maritime tradition.

Before leaving the subject of China's long record of practical innovation, we should point out that the Chinese people also put into practice an invention of a quite different kind: a social philosophy, or way of living together. The individual's goal was to control his or her behaviour, down to the smallest detail of action, emotion and feeling, leading to a respect for oneself that others could recognise and share. This approach to self-knowledge and how it should be applied to create a harmonious society came together in a doctrine set down by a teacher now known universally as Confucius.

Confucius was born at Qufu in Shandong Province in 551 BC into a declining aristocratic family, and he spent his early life in poverty. He was at various times a shepherd and an accountant, and did not begin his lecturing until middle age. His ideas of political reform obliged him to leave his province for fourteen years, but when he returned he opened a school and began to write down his beliefs. He died in 479 BC at the age of seventy-three. That was 250 years before China became a nation, but his ideas had an extraordinary influence on that process of unification, and on the nature of Chinese society ever since. Confucius is buried in a grassy mound in a courtyard full of ancient cypresses in a temple in Qufu. Every day of the year, scores of people from all parts of the world come to stand in front of his tomb, to reflect, and perhaps to rethink their lives a little.

Confucius believed, in very general terms, that if people were comfortable this would make for a more comfortable society, one with long-term peace and stability. His mix of personal and social responsibility suggested a practical way of ordering society—that one should suppress a certain amount of one's individuality for the good of the group; that the educated should lead the uneducated; that rulers should have the loyalty of their subjects, and in turn should help and protect them.

The Confucian ideals were inevitably transformed by his successors, but the key elements of education and service were to hold the Chinese people together for a very long time. Confucius had advocated universal education, and gradually education became crucial in the selection of public officials. In turn, this bureaucracy administered the country under successive emperors according to Confucian principles. The power of the Chinese bureaucracy subsequently waxed and waned, and in later times it became too rigid and stifled the very spirit of innovation that the Chinese people had demonstrated for so long.

But despite the decline during the nineteenth and early twentieth centuries the Confucian-based public service in China provided a continuity which held the country together through more than 2000 years of change—and it still does.

One of China's most renowned men was Confucius, whose ideals of personal integrity and public service have influenced the course of China's history for more than 2000 years. *People come from all over the world to see the grave of Confucius, a green mound in a quiet temple in Qufu, in Shandong Province.*

10.

Pure and Simple

Over the past few hundred years many Pacific societies have seen great change. Some have undergone almost total transformation of their language, economy, culture and in fact their entire way of life. The technology of the European Industrial Revolution enabled the new Western invaders to impose their presence, their economic policies, and in many areas their values, on traditional Pacific societies.

Many collapsed or were swamped. Others decided to catch up with the Industrial Revolution or its space-age successor and become modern industrial nations, with conspicuous success. And though the Pacific club of new manufacturing giants continues to multiply, one above all has adapted so successfully to the ways of the foreigners that it has clearly beaten them at their own game.

That nation is, of course, Japan. Despite the disastrous mistake in trying to emulate the colonial policy of the Western powers, Japan has continued to change and adapt at an astonishing rate. Yet inside this high-tech robot society there is another Japan—one that has not changed in thousands of years, but is still run according to a jealously guarded set of concepts and ideas that the Japanese defend from any contamination. More than most Pacific people the Japanese have retained their original culture and their distinctive way of doing things.

All the parts of this other Japan have a particular character to them. It is visible in the natural logic of a rock or moss garden, the solemnity of the No theatre, the unforced pervasiveness of Shintoism and Buddhism, the inner meaning of the tea ceremony and the very homogeneity of the population. Behind the material excesses of modern Japan there is a simplicity, even austerity, and an adherence to the central and essential elements of existence which set the Japanese and their traditional culture apart from their neighbours. It also underlies their success in abstracting the crucial elements of foreign industrial technology, and with them producing a range of goods which so appeals to consumers that they now

Over centuries of self-imposed isolation, the Japanese evolved their own distinctive culture, in which purity and simplicity became a dominant theme. *This is one of the temples at Nikko, in central Japan.*

189

dominate the markets of the world.

In this brief visit to Japan, one of the great and enduring cultures of the Pacific, we will look behind those striking contrasts to see how the Japanese people have gone about creating a society that sees an aesthetic and even spiritual role for what others might consider the most unremarkable of objects.

The home of the Japanese people is a compact group of islands, and with the exception of their militaristic adventures this century they have spent their entire existence within this clearly defined geographical boundary. The dramatic environmental changes of the ice ages and the huge rises and falls in sea levels have left few traces of the original settlers, but they were clearly part of that northward coastal movement out of Asia, beginning around 50000 or more years ago, which was to populate Siberia and eventually the Americas.

For long periods around that time the lands which now form the Japanese islands were part of the Asian mainland. Archaeological evidence shows that by about 32000 years ago people very similar to those in China were living in the area of Okinawa, which is now an island halfway up the chain that extends from China and Taiwan northeast to the main Japanese islands. Mobile coastal hunter gatherers were active all through Japan by about 30000 years ago. We mentioned earlier the presence of edge-ground stone tools in both Japan and Australia about 25000 years ago, and these may well be indicators of that general and widespread cultural expansion out of Asia into the western Pacific at a surprisingly early period.

One group of Japanese people, the so-called 'hairy Ainu', who were the earliest inhabitants of the northern island, Hokkaido, are so different from other modern Japanese that they have been thought of by some anthropologists as non-Japanese, even non-Asian. Modern research has established quite clearly, however, that the Ainu are not only mongoloid but also closely related to other Japanese. It is possible that they represent the remnant of those early northward-moving people who first explored the Siberian coastline and later North America. The hairiness of the Ainu is probably the result of genetic changes that occurred in a small but relatively isolated group. The steep ridges and volcanic mountains have created many distinctly separate living areas, and this has tended to keep the groups from mixing.

By about 13000 years ago there began to emerge in Japan a sophisticated style of pottery, called Jomon, and there is evidence of an economy based chiefly on fishing and gathering around the coasts, with some hunting of bears, deer and similar game in the forests. This culture coincided with the end of the ice ages, the retreat of the glaciers and the northern ice sheets, and the last great rise in sea level. It appears that people were forced more closely together in the shrinking land areas of Japan, because from about 6000 years ago—when the sea reached its present high level— there is a dramatic increase in the number of habitation sites, including more than 2500 shell middens, around the coastline of the main islands.

Agriculture began in Japan about 6500 years ago, based on the slash-

and-burn shifting cultivation of buckwheat, initially imported from China. Buckwheat is not a grass, and therefore not related to wheat, but is generally regarded as a cereal grain because it fulfils the same dietary role. It originated in eastern Asia—probably China, where its wild relatives are still found—and has many uses. It is high in protein, its grains are fed to animals and poultry, and its flour is used in noodles and pancakes. Buckwheat grows well in poor soils, and this may explain its early spread around the Pacific and eventually into Europe. Over the next few thousand years millet, beans, gourds and peas also began to appear in Japan, and people grew peaches and gathered acorns and nuts.

During the thousand years that led up to the Christian era the Japanese people adopted a generally settled life, with various communities specialising in agriculture or coastal fishing and the gathering of shellfish and seaweeds. But their rather simple agrarian existence began to undergo changes in the first centuries AD, with the growing stream of mongoloid

In their gardens the Japanese express the Shinto belief that the human spirit is in everything. *The garden of the Moss Temple in Kyoto was laid out in the fourteenth century.*

settlers from Korea and China. They brought with them important new knowledge and technology. They introduced rice, and the technique of growing it in flooded paddies. They knew how to make bronze and smelt iron for tools and weapons. The women could weave cloth. These invaders and their ideas were absorbed into the population of Japan without too much upheaval.

In 552 there began a new invasion of Japan, of a quite different kind. The ruler of one of the Korean kingdoms, Paikche, sent to Japan a bronze image of the Buddha and a number of volumes of the sutras, the holy scripts. Two years later he sent a number of men learned in Chinese literature, medicine, calendar-making and music. The Confucian texts soon followed. These spiritual imports rapidly gave way to a rush of practical Chinese technology. Many of the inventions that we described in the last chapter were brought to Japan at this time, sometimes accompanied by teams of practitioners who passed on their skills as painters, weavers, carpenters, potters and metalsmiths.

By the seventh century the Tang dynasty in China was approaching its glittering heights. For the next 300 years, while the West languished through the Dark Ages, China was to be the most powerful, most advanced and best administered country in the world. The effect on its near neighbour was profound. The Japanese adopted the Chinese system of administration and government, and the fundamentals of Chinese philosophy and culture. Chinese style and aesthetic values were fervently taken up in the arts and crafts of silk painting, brocade weaving, ivory carving, lacquer work, pottery, bronze casting and architecture. Chinese urban planning was imported, as was the concept of the formal garden.

Another acquisition was to prove one of the most enduring and influential that Japan was ever to make: the Chinese written language. At this time Japan had no form of writing at all. Having access to Chinese written characters was therefore a great advantage, in some ways, but it led to an enormous problem which has plagued Japan ever since. The Japanese language is polysyllabic, with a number of different sounds making up a word. (This kind of language is ideally represented by an alphabet, but the Japanese had not devised one by the time Chinese writing arrived on the scene.) The Chinese language is monosyllabic, with each syllable or sound being represented by a particular sign or pictograph. This meant that to write a Japanese word of, say, five syllables in Chinese characters, the writer had to find five Chinese signs that sounded like their equivalent Japanese sounds, regardless of what those Chinese signs represented. The results were often ludicrous, and incomprehensible to anyone but the writer. In fact, to be able to write his own language a Japanese had to learn Chinese—a language totally different in vocabulary, grammar and idiom. And to this day, despite enormous efforts to simplify it, the written language is one of the major complications of life for all Japanese.[1]

The history of Japan over the next few centuries—the struggles for power between ruling families and warring factions; the forging of a unified state; the emergence of a warrior class, the samurai, and the

Japanese castles may have a fairy-tale appearance, but they were the centres of power for generations of military rulers. *White Heron Castle at Himeji, begun in the fifteenth century, is perfectly preserved in its original form.*

The armour, weapons and methods of fighting of Japan's feudal warriors, the samurai, still fascinate the Japanese, as well as foreigners. *Alan Thorne holds a keen-edged samurai sword, and behind him stands a suit of samurai armour.*

martial spirit of bushido—is beyond the scope of this chapter. But in 1264 there occurred an event with ominous overtones and long-term repercussions. An envoy arrived from China with a letter from 'the Emperor of Greater Mongolia' to 'the King of Japan'. It suggested that Japan should open 'friendly intercourse' with China, and ended by saying that without harmonious relations 'war is bound to ensue'.

This letter was the manifestation of a huge upheaval in Asia, which had seen the Mongols under Kublai Khan conquer China and set up their own dynasty there. Korea had become a vassal state, and now it was Japan's turn. The Japanese government in Kyoto was frightened, but ignored the letter and other similar demands over the next ten years. Finally, Kublai Khan acted, although his long delay had given the Japanese time to raise an army.

In 1274 a fleet of 450 ships manned by 15 000 Korean sailors and auxiliaries sailed from Korea with 15 000 Mongol, Chinese and Korean troops. The invasion force landed in Hakata Bay on the west coast of Kyushu, the nearest island to the mainland. Although at a disadvantage against the advanced weaponry of the Mongols—including crossbows, catapults and flaming missiles propelled by gunpowder—the Japanese fought fiercely all day, and at close quarters they caused havoc with their terrible swords. The Mongols withdrew to their ships for the night, but a great storm blew up and drove the fleet out to sea, sinking some ships and drowning 13 500 men. The invasion was called off, but the Japanese knew the Mongols would be back, and began building a stone wall three metres high behind the beach at Hakata Bay. Over the next five years they extended it for twenty kilometres. (Part of the wall survives—mostly buried in sand dunes, but in places excavated and restored as a national monument.)

In 1281 the invaders were back, this time with two fleets: one with 50 000 Mongols and Koreans and another with about 100 000 Chinese. Again they landed at Hakata Bay, and again the Japanese resisted fan-

Japan was invaded by the
Mongol armies of Kublai Khan
in the thirteenth century, but
was saved when a 'divine wind',
the kamikaze, blew up and
wrecked the invading fleet. *This
painting of the battle shows the
outnumbered Japanese warriors
challenging the overwhelming
Mongol forces.*

atically. Crowded into the narrow strip between the stone wall and the sea,
the invaders were hampered, and the defenders were able to hold their
own. Their small, agile boats harassed the anchored fleets, setting fire to
ships and sinking boats bringing troops ashore. After fifty-five days the
battle was still in the balance. Reinforcements were hurrying from all over
Japan, but had not arrived, and the superior numbers of the invaders were
beginning to appear decisive. The emperor, nobles, priests and people all
over Japan offered prayers at Shinto shrines for divine help.

As if in answer, a monster typhoon struck the area and blew for two
days. The ships of the invasion fleets were either sunk or driven on to rocks
in the narrow passage. The majority of the sailors and invading troops
were drowned, and altogether some 100 000 men were lost. The Mongols
never came back.

For the Japanese the whole episode had a significance which has never
faded. The typhoon which saved them was seen as a 'divine wind', or
kamikaze. The kamikaze became an enduring symbol of personal sacrifice
and national identity, to be invoked (although sometimes unavailingly) in
times of great peril. But that escape from conquest in the thirteenth
century had many other far-reaching effects on Japan and its people. It
began a period of self-imposed isolation that was to last for 700 years. The
Japanese decided that self-sufficiency, no matter how difficult that might
be, was preferable to the risks of either foreign domination or too great a
reliance on others.

The isolation of the Japanese did give them time to absorb and modify
their Chinese cultural influences, and perfect their own distinctive national
culture, but it also placed severe limits on many aspects of that culture.
Japan is very short of natural mineral resources, which form the backbone
of most civilisations, and so the material side of Japanese life had to be
fashioned from the limited range of natural resources found in the islands:
stone, clay, wood and other plant products such as paper, and a little
copper. To accommodate such constraints, everything had to be stripped
to its essentials.

With little contact outside Japan, new ideas in social and artistic fields
had to be generated internally, and consequently the evolution of social
behaviour, religion and philosophy saw a similar trend towards simplicity
and severity. But instead of leading to sterility, this process made a virtue
of the limits in materials and ideas, and found grace and elegance in the
minimum. So the Japanese created a sophisticated material culture based
on a purity of design, reflected in everything from a rice bowl to a temple.
At the same time they evolved a social fabric and an artistic vision that was
quite unlike that anywhere else in the Pacific basin.

Bamboo is one of the most remarkable plants in nature, and in its
myriad forms is an essential element of life in Asia and Southeast
Asia. But in Japan bamboo has been elevated to an almost
mystical role, and is used to make an extraordinary range of
objects of aesthetic as well as practical value. On the outskirts of Kyoto

there is a living museum to this venerated plant, the Rakusai Bamboo Garden, where we saw something of the vitality and versatility that is the essence of bamboo.

The garden contains specimens of more than fifty species of bamboo, from dwarf forms that make a kind of lawn, to giant stems twenty centimetres in diameter. There are clumps of bamboo whose stems have been enclosed in boxes to make them square. There is black bamboo, and bamboo with alternating yellow and green stripes. One bizarre mutant form is called tortoiseshell bamboo, because each alternating segment is curved like the back of a tortoise.

In the museum there are displays of some of the countless ways bamboo can be used—including a reminder that when Thomas Edison was looking for the ideal material to make the filament for his newly invented electric light globe, he tried hundreds of substances but found that none burned brighter than a carbonised sliver of bamboo. The garden and museum are an indication of how the Japanese combined a limited resource base and an appreciation of natural materials into a style which sets them apart.

To show how the properties of bamboo make it suitable for such an enormous range of manufactured articles we filmed the making of several of them in and around Kyoto, the craft capital of Japan. Here, in workshops that have preserved traditions through many generations, it is still possible to see the imagination and craftsmanship that can transform a length of bamboo into an object of rare beauty and utility.

One of the simplest objects is the most revered of all Japanese musical instruments—the straight, end-blown bamboo flute called a shakuhachi. Its melancholy, eerie tone, combined with the breathiness of the player— which is emphasised, not concealed—evokes the moodiness of Japanese philosophy, particularly the mysticism of Zen Buddhism. In the hands of a master this short tube of bamboo with its five finger holes can encompass three octaves, and an extraordinary range of expression.

We visited a master shakuhachi-maker in his small house and workshop. Stacked neatly in racks below his studio were his raw materials—short lengths of bamboo cut from the bottom of plants grown in special groves on the island of Shikoku, each with its fringe of rootlets in place. (The most famous grove of all was once in Tokyo, but it was destroyed in the fire-bombing of that city in World War 2.) The sections are stored for at least three years to season before being trimmed and cut to length—fifty-five centimetres.

The first stage of making the shakuhachi—the drilling of the five holes, four on top and one below, for the thumb—is straightforward, and often carried out by an assistant. Then come the two processes which test the skills of the maker, and decide the quality of the instrument. The first is the making of the mouthpiece, which involves cutting a narrow slit in one end and gluing in place a wafer-thin sliver of ivory or buffalo horn. This splits the player's breath and sets up the vibrations in the air column which produce the notes. There are no fixed dimensions or angles for this; it is a matter of experience. The most critical process is the final tuning.

Bamboo is an essential element of life in Asia, and in Japan it has been given almost mystical status. *The Rakusai Bamboo Garden in Kyoto grows more than fifty species, including this tortoise-shell variety.*

The final tuning of the shakuhachi is done by applying thin layers of plaster and lacquer to the inside of the barrel. *This master flute-maker in Kyoto spends up to a month on each instrument.*

One of the most revered musical instruments in Japan is the bamboo flute, called a shakuhachi. *The finest of these, made to order for leading musicians, may cost as much as $30 000.*

This is done by applying thin coats of plaster and lacquer to the inside of the flute with a long, narrow spatula. Each layer takes several days to set and dry, and between each one the maker must play the flute and check the tuning, until he is satisfied.

The making of shakuhachi, restricted now to a few craftsmen, is precise and unhurried. A flute-maker usually works on several instruments at the same time, and each one may take a month or more to finish. The finest shakuhachi are made only to order, and may cost tens of thousands of dollars each.

Not every Japanese plays the shakuhachi, of course, but everyone uses the bamboo-handled calligraphy brush, a tool that is central to Japanese life. The first brushes were imported from China with the written Chinese language, more than 1000 years ago, and for a long time the craft of brush-making was controlled by the Imperial court. Like the shakuhachi, brushes are deceptively simple objects, but they, too, are still made in Kyoto by craftsmen who hand the tradition down from one generation to another.

The key to the performance of a good brush lies in the choice and blending of the hairs in it: sable, badger, goat, sheep, horse, deer, weasel, rabbit, racoon, or bear. The brush-maker selects for quality and carefully grades the hairs into five lengths, to provide variable flexibility from the base of the brush to its fine tip. Then he rolls them in ash to remove any trace of natural oils, which would repel ink. The brush-maker divides and mixes, sorts and arranges the hairs according to long experience, before separating them into small bundles for tying and inserting into their bamboo handles.

A finished brush may look like a foreign mass-produced product, but its price suggests otherwise. The demands placed upon it in use are also quite different. It must respond, almost like a living thing, to the hand of the writer. Much more than any form of Western writing, Japanese calligraphy is both art and technique. It embraces not just information but style and personality. It combines a technique acquired over years of practice with a spontaneity of feeling. An ink stick made of soot from burned rapeseed oil mixed with animal glue and perfume, is ground on an ink-stone and mixed with water. The brush is loaded with ink, and the kanji (characters) executed with a flourish. The course of the line must be confident and unerring. It may vary in strength, but never hesitate. And all that it expresses must come together at the one moment. With Japanese calligraphy there are no second thoughts.

In Japan there are a hundred other simple bamboo creations, but one above all is uniquely Japanese. It is made from a single short piece of bamboo and a cotton thread, and it is called a chasen. The word means tea whisk, and it is used to mix the powdered green tea in the tea ceremony. Chasen-makers follow an ancient and honourable tradition that in some families goes back seven or eight generations. The person we filmed at work, near Nara, is chasen-maker to the emperor.

The chasen begins as a short tube of bamboo about fifteen centimetres long, with a joint about two-thirds of the way from one end. It is this longer section which is turned into a delicate cage of fine bamboo filaments. The craftsman first uses a small sharp knife to make sixteen radial splits around the end of the tube. Turning the blade, he then splits these segments into inner and outer rings. The inner ring is broken out, but further splitting of the outer ring continues, until it is transformed into a ring of up to 120 fine bamboo slivers. Each of these is shaved finely and curled inwards, after being softened in hot water. A cotton thread is wound in and out between the filaments to separate them and to enhance the appearance of the chasen. When finished, it looks like a delicate bamboo cage. The chasen made by the craftsman we filmed are not only used in the tea ceremony but also kept as objects to admire.

The Japanese tea ceremony is an excellent example of the way in which the Japanese have invested meaning and even spirituality in something that relies on the utmost simplicity—what others might consider a thoroughly mundane event. To most foreigners the tea ceremony, with its deliberate, formalised routine, in which the objective is not actually to enjoy drinking tea, may seem quite pointless. For the Japanese, however, the whole experience is charged with meaning. Today, with the shockwaves of Westernisation spreading ever more widely through Japanese society, participation in the tea ceremony remains an extremely important source of reassurance about traditional values.

The ostensible purpose of the tea ceremony is a gathering of friends with shared artistic tastes, conducted according to prescribed etiquette and in quiet surroundings. The friends gravely discuss an admired piece of pottery, or a painting on the wall, or a poem inscribed on bamboo, while green tea is prepared according to strict rules and small sweetmeats are consumed. Expressed baldly in that way, the legendary tea ceremony may indeed seem mundane. But as we filmed it in Kyoto we became aware of a remarkable hidden agenda.

The tea ceremony began in the fifteenth century and grew out of the Zen

The chasen, or tea whisk, is an essential element of the tea ceremony, and is made by progressively splitting and shaping a single tube of bamboo. *Chasen created by famous makers are treasured as objects not only to use but also to admire.*

There are few chasen-makers left in Japan, and their products are eagerly sought after. *This master craftsman is chasen-maker to the Emperor.*

habit of drinking tea as an aid to meditation. It was a time of turmoil and trouble in Japan, when many beliefs were under threat. The appeal of the peaceful interlude offered by tea drinking spread outside Zen circles, and gradually the ceremony became formalised and widely practised in all ranks of Japanese society. Tea masters who presided over the proceedings established strict rules. For more than 400 years now its spare ritual has not varied, except for some extravagant interludes in the sixteenth century.[2]

The tea ceremony is held in a room which is traditionally a 'four and a half mat' room. This refers to the size of the prototype tea room, measured in mats, in the Silver Pavilion (Ginkaku) in Kyoto. This was built beside a lake in the fifteenth century by the Shogun rulers to express the architectural ideal of simplicity and 'concealed beauty', which must be summoned forth by the trained taste of the connoisseur. The mats were of fixed dimensions, and four and a half mats made a room about three metres square. At one side of the Ginkaku tea room there was a small, slightly raised alcove, in which a single treasured object was placed to form the subject of discussion. (This alcove originally contained a shrine, but was later incorporated into every Japanese house as a decorative feature—an indication of the pervasive influence of the tea ceremony on Japanese domestic arrangements.)

The tea master kneels in one corner of the room, half-turned away from the guests—usually two—who kneel beside the alcove. He is dressed in a black robe, and they in their best kimonos, with a special folding fan tucked into each. The fan is different from the usual Japanese folding fan and is used only at tea ceremonies, and then only as a 'tray' for passing sweetmeats. The tea master has water boiling in a cast iron kettle of particular design on a dish of charcoal, a decorated tea caddy containing powdered green tea, a bamboo spatula for measuring out the tea, a bamboo dipper for ladling out water, a bamboo chasen for whisking the tea, and two handsome pottery drinking bowls. Each item has a particular place, and after being used it is replaced precisely.

While the guests politely discuss the object of their attention, the tea master slowly and carefully places a small quantity of tea in a bowl, adds a little water in a precise and ritualistic manner, and whisks the mixture into a thick froth. In the meantime the guests ritually taste small titbits made of sweetened bean paste and bean jam (omogashi). The tea master turns and places the tea bowl in front of one guest, turning it so that the correct side is in front, and bows.

The guest carefully moves forward on his or her knees with three slow, gentle movements, takes up the bowl, and retreats in the same careful fashion. Both guests then bow to the tea master, and the one with the bowl rotates it to the correct position, and ritually sips the green mixture. Meanwhile the tea master begins to prepare tea for the other guest. The entire ceremony may take an hour or more.

The significance of this deceptively simple ceremony lies in two areas, both important to the Japanese. The artistic appreciation, the intellectual conversation, the controlled gestures and attitudes—all have Buddhist

The tea ceremony, although outwardly simple, has deep significance for the Japanese. *The two guests, wearing their best kimonos for the occasion, are about to receive their green tea from the tea master.*

connotations. But these rituals, the precise ordering of the manner in which things are done, are more than simply traditional forms of behaviour. They define the course of Japanese life. These stylised and very formal manners ensure that everyone at all times knows exactly what each must do. They provide stability in a turbulent world, and the reassurance of belonging to an extremely homogeneous society. Taking part is a reinforcement and an act of security.

The second appeal is easy to understand. Japan is perhaps the most crowded country in the world. There is a shortage of psychological as well as physical space. There are great pressures on the individual and little room to manoeuvre. The tea ceremony provides an acceptable and approved opportunity to withdraw from daily cares, and to leave for a time the tensions and disorders of the world outside.

In the headlong rush of the Japanese to catch up with the West, and in some aspects of life to embrace Western values, many traditional Japanese attitudes are changing—but in some areas the Japanese are resisting change. The traditional house is one of the last bastions of simplicity in design and honesty, in the use of a limited range of materials.

The simple wooden beams and posts, the sliding paper doors and windows, the rendered clay walls, the reed and straw flooring, all are visible. Nothing is hidden here, and there is never a facade as there can be in the West. Yet this ancient style has influenced many so-called 'modern' Western ideas of architecture and interior design—the exposed structural elements, the modular units of standard size, the recessed handles on doors and windows, even the bamboo and rice-paper lanterns.

The rooms of a Japanese house are few and sparsely furnished, often with nothing more than a low table. There are no chairs or beds, because the shortage of space precludes the setting aside of a room just for sleeping. The sleeping mat or futon is rolled up and put away behind a sliding panel during the day. The floor is the key element in the room, and its special nature is responsible for one of the most widely observed customs in Japan: the removal of shoes before entering the living quarters of the house. What makes the floor special are the close-fitting tatami (mats) which cover it.

Tatami have been in use in Japan for more than a thousand years, and they are the heart of the Japanese house. Life evolved on the tatami—sitting, eating, sleeping, creating life and departing it. The cool, smooth, springy, tightly woven surface offers rest for the eyes and a comfortable resilience for the feet. (And that is why shoes are removed, because no footwear, not even slippers, must touch that immaculate surface.)

A tatami consists of a thick pad of compressed rice straw about five centimetres thick, covered with a stretched, woven mat of pale green igusa reeds, and the edges bound with cotton cloth. The mat should last for at least fifteen years or more, provided it is re-covered occasionally. The mats come in only three sizes in all Japan—the largest, in the Osaka-Kyoto area, measures 192 by 96 centimetres. Houses are built around the dimensions of a given number of tatami. In the average house, rooms range in size from 'three tatami' to 'ten tatami'. In temples, palaces and castles there may be as many as a hundred tatami in one room.

It is therefore possible to understand how the tatami, roughly two metres by one metre, became the basic unit of Japanese architecture, and to some extent of social life. It represents the living and sleeping space of one person. In a country which had little *lebensraum* to start with, and has progressively lost more with the urbanisation of the past 200 years, this explains the fascination of Japanese people for the wide open spaces of countries like Australia and America.

Even though these highly functional and cost-efficient houses are steadily being replaced by modern structures, particularly in cities, there is one tiny room in every Japanese house where tradition and deeply ingrained habits still prevail, and that is the kitchen. The Japanese preferences in food, and their style of preparing it, have survived the attractions of Westernisation better than many aspects of their lives. The two staples of the Japanese diet are still, as they have been for thousands of years, rice and fish, and in adherence to the principles of purity and simplicity they invariably eat the rice plainly boiled and the fish raw.

The Japanese reliance on marine foods for their protein grew initially out of necessity, for their steep, crowded islands were never suitable for raising the large meat-producing animals on which other cultures depended. This is what makes the fish markets in the larger centres of population so important, a bond between the urban Japanese and the sea that they can no longer harvest themselves.

In the Kyoto fish market, before dawn one morning, we filmed perhaps

The reverence of the Japanese for the pure and simple is reflected in the austere elegance of the Japanese house, in which natural materials are used without decoration or concealment. *The tatami (mats) of compressed rice straw, covered with woven reeds, are a particularly important element, and no footwear, not even slippers, may touch them.*

In the Japanese diet, which is
dependent upon sea food, the
most important ingredient is
fresh, raw tuna. *A restaurant
supplier slices tuna bought at
the auction held every morning
at the Kyoto fish market.*

The combination of the
Japanese taste for raw fish and
their technology can be seen in
the sushi automats now
appearing all over Japan. *The
portions of sushi—all one
price—are carried on a
continuously moving belt past
the customers, who are charged
by the number of plates they
stack up beside them.*

the largest and most varied array of marine foods ever assembled under
the one roof. Brought by the huge Japanese fishing fleets that roam the
Pacific, as well as the thousands of bays and estuaries around the Japanese
islands, the display is overwhelming. There are countless species of fish
and scores of varieties of every other class of sea creature—crustaceans,
molluscs, sharks and rays, starfish, octopuses, seaweeds and other algae.

There are large slabs of red whale meat, too, as Japan fights to retain the
right to eat these great marine mammals. But unquestionably the most
important single marine product for the Japanese is tuna. Every day the
centre of attention among the buyers is the tuna auction which begins
sharp at 5.20 am.

The tuna on offer are laid out in neat rows, each on its own slab, cleaned
and trimmed, blue-black skin glistening under the lights. The tuna come
from far and wide, from the north Pacific to the Antarctic, even from the
Indian Ocean. But so insistent are the Japanese on freshness that freezing
of the catch is not permitted. The tuna are packed in ice as soon as they are
caught, and flown, at whatever expense, to the markets in Japan.

As the buyers inspect the day's offering, they run a thumb across a small
flap cut near the tail to judge the quality of the flesh. That is all the
handling allowed. No buyer will bid for any tuna that is marked in the
slightest. Some Australian tuna fishing boats catching for the Japanese
market have had to line their decks with carpet to prevent bruising of the
tuna as they are heaved aboard by the men working the poles and lures in
the stern.

The bidding, when it begins, is fast, knowledgeable and highly com-
petitive. One bidder holds up his cap and makes a secret sign inside it with
his fingers, one that only the auctioneer can see. There is considerable
prestige in securing the best fish of the day, and this is usually achieved by
a wholesaler who supplies restaurants.

The auction prices reflect the almost mystical value put on the best tuna
by Japanese customers, so that even the top restaurants can rarely afford
to buy a whole fish. We watched one middleman carefully slice the best
tuna of the day, cutting blocks from the different parts of the fish, which
fetch different prices. For the whole tuna he paid seventy dollars per
kilogram. He told us that the best cut, the pale flesh from near the fin, will
cost the restaurant customers who eat it later that day the equivalent of
1500 dollars per kilogram.

The remarkably efficient distribution system that covers the compact
Japanese island chain ensures that marine products reach the markets
within hours of being caught, and virtually everything sold at dawn is
eaten the same day. The freshness guarantees the purity the Japanese
insist upon, and the simplicity they favour comes from eating their sea
food raw.

So important is sea food to this nation that there has grown up a whole
set of conventions and rules about how to prepare and serve it. The slicing
of raw fish for sashimi or sushi is not just a matter of cutting large pieces
into small portions. Each fish is skinned and sliced in a particular way, and

to facilitate this delicate operation the Japanese have no fewer than thirty-one different knives just for slicing fish. Some fish, including tuna, globe-fish, eel, and mackerel, have a special knife all to themselves. The angle of the cuts to the grain of the flesh, the size and shape of the pieces from different fish, the juxtaposition of the various kinds of fish on the serving platter—all have to be correct, if the presentation is to be acceptable.

The bond between the Japanese people and the sea goes back to their earliest settlement of these islands, when all people were hunters and gatherers, and harvested the sea for a living. The Japanese still harvest the sea in every possible way, from the women who collect seaweed along the shores of the Inland Sea, to the modern fishing fleets with their helicopter spotters and sophisticated sonar fish-finding systems, far out in the Pacific. But there is one link with those ancient traditions which expresses that continuity with the past better than anything else—the ama or diving women, who forage on the sea bed for abalone and other sea food in the shallow coastal waters of Japan and Korea.

The ama have a known history that goes back at least 1500 years. Even as recently as twenty years ago there were still 30 000 women in the diving industry, but today the numbers are dwindling. Most are in Korea, though there are still several thousand along the south coast of Japan. And that is where we went to see them at work.

Near the small village of Goza on the Shima Peninsula we got up early, as the ama do, and followed them on to the small motor boats they use to reach the abalone grounds, just off the rocky coast. The women were laughing and chattering as they prepared for their day's work, and the most startling thing about them was their age. Many of them are more than sixty years old, and most are in middle age. Fewer young women seem to want to learn this arduous trade.

The ama or diving women of Japan and Korea, who search the sea bed for shellfish and edible seaweeds, maintain a tradition that goes back for thousands of years. *The ama keep their catch—like this sea urchin—in a floating basket until they return to their boat.*

Until quite recently the ama dived naked, wearing only a pair of goggles made of naturally transparent material, mica. Today they wear cotton or rubber wetsuits, and their goggles may have prescription lenses built in. They use no form of breathing apparatus, not even a snorkel, but rely on their remarkable lung capacity. Medical studies have shown that over a few years each ama makes a set of physical adaptations, which include a slowing of the heart rate and a restriction of capillary circulation in the skin to cut down heat loss in the water. The sea off southern Japan is slightly warmed by the Kuroshio Current from the south, but several hours in the water, even with a wetsuit, is all the body can stand.

When we reached the diving grounds the boats cruised in circles, and the women began jumping overboard. Each took a bucket fixed inside a rubber flotation ring, a short metal lever, and a small waist net attached to a belt. The bucket floats on the surface, attached to the ama by a long cord, while she dives to the bottom and collects shellfish, sea urchins and edible seaweeds. The lever is used to prise abalone off the rocky bottom. Returning to the surface, the ama swims to her bucket and transfers the catch from her waist net. She rests for perhaps thirty seconds, giving a peculiar whistle as she expels the build-up of carbon dioxide from her lungs, and then dives again.

The ama we were watching were diving in about ten metres of water, but they can work in depths of up to twenty metres. They swim rapidly to the bottom, spend about half a minute searching, and surface quite quickly. In shallow waters they can stay down for up to two minutes if they have to. Including their rest, and depending on the depth they have to dive, they can make up to thirty dives per hour. The time they spend under pressure is too short for nitrogen to be forced into the bloodstream, and therefore they are in no danger of suffering the bends.

The ama in Japan dive for about three hours each morning and again for three hours in the afternoon, but now only in summer; the Korean women still dive all year round. This routine, six days a week, enables the women of Goza to feed their families, and sometimes sell a little surplus in the local market. Watching these strong, cheerful, independent women it was hard to resist the conclusion that while they could certainly make a living in some less arduous way they really love their life in and under the sea.

We went ashore on a small island with the ama at midday, to a group of huts behind a deserted beach. They stripped off their wetsuits and hung them up to dry, while they gathered round a blazing fire in one of the huts to eat and to warm their bodies.

As they talked and laughed, and eventually sang a haunting little song of their life and the sea, the ama seemed to us to personify many aspects of the character that has formed the Japanese people over the past few thousand years. They maintain a close link with nature, and rely on a minimum of technology to win from it what they need. They are very conscious of the balance between self-reliance on the one hand, and the needs of the group on the other. These women are cheerful and friendly to strangers, but also physically strong and tough-minded—qualities which helped to win and keep the independence that Japan so desperately fought for in the past, and which surely go a long way to explaining the phenomenon of the new Japan.

The ama, who are nearly all in their fifties and sixties, represent the spirit of self-reliance and independence that has been such a feature of Japan's history. *The women are warming themselves during their midday break from diving.*

11.

Flaming Arrows

W hile all this diversification of human society was taking place on the western side of the Pacific, across the other side of that vast ocean the explorers who pushed north into Siberia during the ice ages were making their greatest territorial conquest: the occupation of the Americas.

As we saw in an earlier chapter, that process began at about the same time as the first ocean voyagers were island-hopping south and east across tropical seas into Melanesia and Australia. From Siberia, little bands of hunters skirted round the top of the Pacific basin, following the coastline of Beringia towards North America. The routes they took and the date of their first arrival in the new continent are still matters of great debate. But what seems now beyond dispute, based on archaeological evidence from many parts of the two Americas, is that from a fairly narrow entry in the northwest corner, some time before 40 000 years ago, those Asian migrants spread out and gave rise to an astonishing variety of societies and cultures, right down to the tip of South America.

One of the first areas where these colonists paused and put down roots, defining territories and learning how to make a living as Americans rather than as Asians, was the northwest coast of the continent. The Pacific northwest, from Alaska down the coast of British Columbia to Washington State, just beyond the border between Canada and the United States, formed then as it does now a clearly defined and quite distinctive environment. It provided those first settlers with a bounty of resources matched in few other places in the world.

This picturesque coastal strip, with its forested islands and mountains coming right down to the sea, its sheltered bays and narrow, rocky channels, is more than 2000 kilometres long but no more than 300 kilometres wide, backed by high mountain ranges which cut it off from the interior. This barrier blocks the moisture-laden onshore winds from the warm Japan Current which flows down the coast, resulting in an annual rainfall of more than 4000 millimetres. The combination of high rainfall

The original settlers of North America were swamped so quickly by the tide of European occupation that the richness and diversity of their cultures were gone almost before we knew they existed. *A few of the great totem poles still stand at Ninstints, an abandoned village of the Haida people in the Queen Charlotte Islands, off the coast of British Columbia.*

and temperatures which rarely fall below freezing created dense forests of giant redwoods, firs and cedars, inhabited by a great variety of animals, from birds and rodents to deer and bears. The seas were alive with whales, seals and otters, and the rocky shores covered with oysters, mussels and abalone. There were 200-kilogram halibuts to be caught, as well as six-metre-long sturgeon, monster cod and huge shoals of candlefish—so oily that they could be literally set alight to burn like candles. Salmon swarmed up the rivers to spawn in such numbers that they could be scooped out in hand nets.

The abundance of timber became a mainstay of the material culture of the northwest people, and helped their development as the most original and accomplished wood carvers in all the Americas. Red cedar made wonderful canoes, and also split easily into planks for house-building. Soft yellow cedar, which does not split, was ideal for carving into masks, bowls and implements. Flexible yew made bows and harpoon shafts. And the mighty trunks of fir made house beams and the towering totem poles, carved with mythical animals, which became the most striking art form of the northwest coast.

Because of their comparative isolation from the rest of the continent, the people of the northwest retained their culture longer than many others after new colonists from Europe had begun to arrive in North America. From reports about the northwest coastal tribes in the eighteenth and nineteenth centuries it is possible to obtain a picture of the colourful existence that people had enjoyed on that wild and beautiful coastline for thousands of years.

There were six or seven major tribal divisions, and four major language groups, but the life-style of all the coastal tribes was very similar. The people lived in large wooden houses up to twenty metres long by ten metres wide. These were built around massive fir logs driven into the ground, with the roof supported by huge beams, cut and shaped with stone tools. A house was occupied by a number of families, each with its own fire

The people of the northwest coast of North America were great fishermen and hunters of the sea, in canoes made from the trunks of red cedar or fir. *The Haida people of Skidegate, up the Canadian coast from Vancouver, recently made this canoe in traditional style.*

hearth and living space. Like their Siberian ancestors, the people sometimes dug out the floor area to create a huge sunken fire pit in the centre,
with terraces leading down to it, and an opening in the roof to let the
smoke out. The beams and planks forming the front of the house were
decorated with images of mythical and real animals—ravens, bears,
whales, dolphins. Ten or twenty of these striking buildings formed a
village, usually facing the sea. In front of the village the people erected
their great totem poles. These were up to twenty metres high, and
strikingly carved and painted.

The people were primarily hunters and fishermen in seas that were
rough and stormy, and one of their greatest skills was boat-building. They
made boats up to twenty metres long from a single trunk of cedar or fir.
They felled these forest giants—some more than two metres in diameter
and sixty metres tall—by building controlled fires around the base. They
hollowed out the log with stone adzes and small fires, and chipped away at
the outside, until they had a huge dugout canoe. The final step, of
broadening the beam to make the boat more stable, used a quite ingenious
technique. They filled the canoe with water and dropped hot stones into it.
As the steaming hot water softened the timber the two sides were forced
apart with thwarts or spreaders, which later became seats for the paddlers.

Summer was the time for catching candlefish in the channels and
salmon in the rivers. The candlefish were too oily to be smoked or dried
and were boiled to obtain their oil, which was a valuable trade item with
inland tribes. Salmon were netted in huge numbers as they fought their
way into the rivers, and most were split and sun-dried for winter use.

Winter was the wettest season, and time for intense social activities in
the big houses. The most important event was the potlatch, a word which
meant 'to give'. A potlatch was held whenever a tribal chief chose to or was
obliged to celebrate a death, wedding, or similar event. The occasion was
designed to reinforce the social standing of the chief and his village, by
feeding large numbers of people and giving away valuable status symbols
such as canoes, shell ornaments, cloaks woven from mountain goat wool
and bark fibre, baskets, or pieces of beaten native copper. In this society
the capacity to give things away, rather than the accumulation of material
wealth, was the basis of community respect and honour.

Up the coast from Vancouver, in the rugged wilderness of the Queen
Charlotte Islands, we visited the sites of villages once inhabited by the
Haida, one of the most powerful of the northwest tribes. At Ninstints,
beside a quiet beach on a sheltered channel, the collapsed ruins of great
houses and the decaying totem poles are mute reminders of the society
that was once so active here, and the disease and death which caused it to
be abandoned more than 100 years ago. The forest is steadily reclaiming
the ruins, covering the fallen beams with a soft green blanket of moss.

There was once a community of 300 people here, in twenty big houses,
their whaleboats and canoes drawn up in rows on the beach. A dozen of the
great totem poles still stand, scarred by fire and decay, some upright,
others leaning forward, their huge carved eyes staring at the ground. The

In a revival of their traditional wood-carving skills, the Haida people are once again creating the great totem poles. *This carver in Skidegate is making a pole for display outside the community centre.*

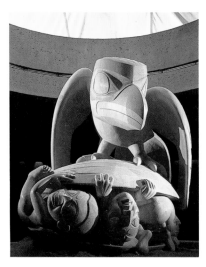

This massive sculpture in yellow cedar by the Haida master carver, Bill Reid, tells the legend of the raven who found the first men emerging from a clam shell on the beach. *The statue, which stands nearly two metres high, is in the Museum of Anthropology of the University of British Columbia in Vancouver.*

figures on these poles—bears, whales, frogs and thunderbirds—record the natural world that surrounded the Haida and the mythology which ruled their lives. In their style they are also a clue to the world of Siberia that the ancestors of the Haida left behind in Asia, all those thousands of years before.

In the modern settlement of Skidegate, the surviving Haida people live on, clinging precariously to their traditions after the loss of so many of their elders and craftsmen in the epidemics of introduced diseases. But the famous wood-carving skills are reappearing, most practically in the construction of two cedar canoes—the first to be made on the coast for more than eighty years. One was big enough for forty people, and was paddled down the coast to Vancouver and back, a voyage in open waters of more than 800 kilometres. The carving of totem poles has begun again, too, and some of the most striking can be seen in the grounds of the Museum of Anthropology of the University of British Columbia in Vancouver.

The moving spirit behind the revival of Haida pride in their artistic traditions is Bill Reid, a master carver who led the team which made the great canoe. More recently, Reid stretched his imagination and his materials with a monumental sculpture in yellow cedar, three metres high, which was unveiled in the museum in 1980. 'The Raven and the First Men' shows a great raven sitting on a partly open clam shell. Forcing their way out of the shell are human figures, infants and old men, seeking life. The raven is the most important creature in the mythology of the northwest coast, and in the Haida legend the raven finds a clam shell on the beach, with the first humans crawling out of it. The raven teaches these naked beings all the things they will ever know, and then goes off to create all the other creatures in the universe.

Another tradition that is undergoing a striking revival is the carving of dance masks. These were used in the ancient ceremonies of the Tlingit, Tsimshian and Kwakiutl people. They are carved in wood to represent stylised creatures of many kinds, and painted in bold colours—black, red, green—and sometimes trimmed with feathers or animal furs. The best ones being made today are imbued with all the mythological spirit of the originals, and display carving skills of great assurance. Of similar high quality are the striking silk-screen prints of traditional figures and animals, in bold black and red forms, that are being produced by northwest coast artists.

The original success of the northwest coastal people in learning to use their rich and beautiful, but often cold and dangerous, environment was a good start to the colonisation of the North American continent. It was soon to be repeated in scores of new and different environments, as the early colonists spread across the virgin territory. It is impossible for us to give a comprehensive description here of all those societies and the amazing inventiveness of their art and culture, but a look at the achievements of several distinctive groups does convey some idea of that diversity—a richness that was overlooked in the rapid European occupation of North America.[1]

It is all gone now, but further south, in what is now California, there was a continuation of the life-style of the northwest coast, adapted to the warmer climate. Again, being linked to the coast, the people here were excellent boat-builders and fishermen. They used redwood canoes to hunt seals and otters along the coast. Some also revived the ancient Asian practice of fishing at night, using flaring torches to attract fish to the canoes, where they could be speared and netted. (Such methods are still in widespread use in Indonesia.) Out of the giant redwoods they carved large flat-bottomed canoes, up to twelve metres long and capable of carrying twenty-five paddlers. The Chumash, who lived on offshore islands where large trees were scarce, made canoes of planks lashed together.

On the inland lakes some of the California tribes used craft made of bundles of reeds, very similar to those still in use on Lake Titicaca in the Andes, and on the coast of Peru.[2] These could be very large—the Yurok trading canoes were up to fifteen metres long. They were made from three

These striking dance masks are examples of the revival of the art of carving on the northwest coast of North America. *The masks form part of the magnificent collections in the Museum of Anthropology of the University of British Columbia.*

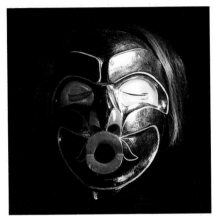

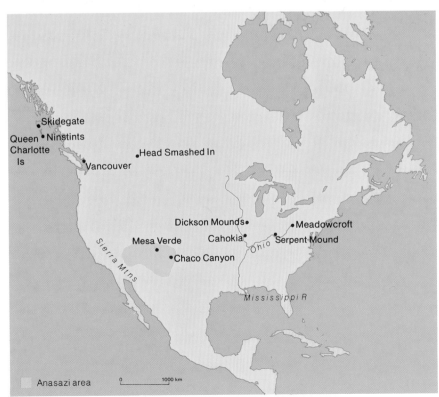

209

large bundles of reeds, one for the bottom and two for the sides, bound together with willow stems and turned up at both ends. Such a craft was home for two families for up to ten days, as the men poled them up streams and around lakes to attend gatherings where goods were traded. The vessels, being highly perishable, have not survived, but early drawings of them suggest that a fleet approaching across a lake must have presented an impressive sight.

Another ingenious watercraft used by the people of the San Joaquin valley in the southern Sierras in California was the basket boat, obviously derived from their skills in basket-weaving. This was round, with a flat bottom, and about a metre and a half in diameter. It was made entirely from reeds, woven so tightly that it required no waterproofing. (Waterproof baskets were even used for cooking food in California; the heat was applied by dropping hot stones into the basket.)

One interesting practice widely used in California arose from a set of environmental conditions that produced an exactly similar reaction by hunter gatherers across the other side of the Pacific, in Australia. This was the regular autumn firing of the pine and redwood forest undergrowth, the oak-studded woodlands, and the dense chaparral or brush. The objectives were to prevent the forests from being invaded by dense and unproductive brush, to make the landscape easier to walk through, and to promote the growth of new grass that would attract grazing animals and make them easier to hunt. This practice also prevented the build-up of fuel in the forests, and therefore reduced the risk of extremely fierce fires. Such holocausts destroyed large trees that were valued for their timber, their nuts, and the habitats they provided for animals. (There are no native Californians still burning in the traditional way, and the consequences can be seen in the catastrophic bushfires which are such a problem in that populous state. The parallel between North America and Australia in the effects of aboriginal burning—and in fact the widespread modification of the landscape by hunter gatherers—is discussed in our final chapter.)

It is still not clear when the first Americans crossed the great barrier of the Rocky Mountains, although there is some evidence from the eastern side of the continent that it must have been well before 20000 years ago. This evidence comes from a cave at Meadowcroft in Pennsylvania, on the Ohio River near Pittsburgh. Here, in various levels dated to between 12000 and 19000 years ago, archaeologists have found finely worked stone tools, charcoal hearths, and the bones of deer, chipmunks and squirrels that were apparently hunted in the mixed forests of deciduous trees and conifers that grew in the area during the ice ages.

Apart from this site, however, there are very few traces of people east of the Rockies during the ice ages. This is extremely puzzling, since the many well-dated sites in Central and South America make it certain that at least a substantial corridor along the Pacific coast of the Americas was in use by humans by about 40000 years ago. One possibility is that the earliest people east of the Rockies were not hunters, whose stone points would show up in habitation sites, but perhaps gatherers of plant foods, which leave little trace. But whatever the reason for the mysterious absence of people early on, the scene changed very dramatically about 12 000 years ago, with the appearance on the great plains of North America of the big-game hunters.

Today there are fifteen species of large animals in North America, but when the ice ages ended there were more than seventy—mammoths, mastodons, sabre-toothed cats, giant bears, bison, camels, horses, and huge slow-moving ground sloths that were six metres long and weighed three tonnes. One estimate, based on the profusion of bones of these now extinct species, suggests that there might have been 50 to 100 million large animals north of Mexico. Their disappearance during a relatively short period about 11 000 years ago is now generally attributed to human hunting. The people responsible are thought to be the Clovis hunters, so named because of the large, finely made stone hunting points found at Clovis in New Mexico, in sites dating from that period.

This process of extinction has been called the 'ice age overkill', although there are some who believe that environmental factors, such as severe climatic change, were as influential as human hunting. It may be that climatic changes put certain species under such environmental stress—through loss of habitat, for example—that even a few hunters could have pushed them over the brink into extinction. Whatever the reason, the disappearance of so many large animals must have forced a considerable change of life-style on the Clovis hunters and others living on the prairies. Many tribes were forced to specialise on one animal that did not disappear, and they succeeded in making it the basis of their existence. That animal was the buffalo.

The Navajo still display the strong character of their once free-ranging tribal warriors. *This man is more than eighty, and has a large family of children and grandchildren.*

The area between the Rocky Mountains and the Great Lakes was a sea of grass, maintained in some areas by regular burning to prevent the incursion of woodland, as in California. There were antelopes, bears and several kinds of deer on the plains, but the dominant animal was the buffalo, in herds that numbered millions. To hunt these big, powerful animals the plains people had to develop methods which utilised the natural behaviour patterns of the buffalo, since they could not hope to run them down on foot. (The original horses in North America became extinct at the end of the ice age, and were not seen again until they were introduced by the Spaniards.)

All across the plains there are sites which illustrate how the hunters worked. Among them are ravines choked with buffalo bones, and hundreds of discarded flint spear points and butchering knives. One site contains more than 40 000 bones of buffaloes apparently trapped in winter by hunters who stamped down the snow at the entrance to a gully to make a glassy chute. Animals seeking feed or shelter were stampeded to their deaths below. A more efficient and spectacular operation was the buffalo jump, practised all the way from Alberta down to Texas. We visited one of the most famous jumps, called Head Smashed In, 100 kilometres south of Calgary and within sight of the Rockies.

At Head Smashed In the Piegan people had worked out a most effective method of killing buffaloes in very large numbers. The site was a high sandstone cliff facing the plains. Back from the clifftop there was a shallow, well-watered basin where herds of buffaloes often gathered to feed. The Piegan hunters would creep up from behind the hills and form a huge arc and begin slowly moving the herds down a valley towards the cliff. Cairns of stones and piles of brush on the hillside helped to guide the buffaloes in the right direction. As the herd neared the cliff the hunters increased their pressure, finally leaping up and shouting to raise the animals to a full stampede. There was no stopping at the cliff edge, and the herds plunged over in hundreds. At the bottom of the cliff other people, including women, butchered the dead and dying animals. As well as fresh meat, they obtained fat and meat for storing in the form of pemmican, hides for making tents and clothing, bladders and intestines for containers, sinews for sewing and binding, and bones for tools and utensils. Head Smashed In was used in this way for thousands of years, and excavations show that the grassy slope at the foot of the cliff is actually a bed of buffalo bones more than ten metres deep.[3]

The plains people were totally dependent upon the buffalo for their existence. When in the nineteenth century the great herds were decimated by European shooters (3.7 million were killed in three years, from 1872 to 1874), it was a greater threat to the tribes than their fights with the railroad-builders or the cattle-raisers. By early this century the herds had been reduced to a few thousand animals, and whole tribes, and their cultures, had simply disappeared from the Great Plains.

One area of North America which saw a great diversification and evolution of cultures was the arid southwest, with its spectacular canyons and buttes, weathered from the many-coloured sedimentary rocks. Today the southwest is harsh enough, with hot dry summers and bleak winters. During the ice ages it was even more inhospitable, and yet there were people living there well before that period of climatic extremes came to an end.

The desert cultures of North America were very similar to those in Australia, dependent upon an opportunistic and highly mobile life-style. The range of animals to be hunted, very limited by comparison with more favoured environments, included the occasional bison but more often antelopes, deer, sheep, rabbits, snakes and lizards, and small burrowing rodents. Plant foods were scarce, and usually found by following the pattern of the patchy and infrequent rainfall.

Then came a major change. Further south, in what is now Mexico, plants had been domesticated, including maize, or corn. When the knowledge of how to grow corn came north, about 2000 years ago, the cereal quickly became a staple. Corn provided a reliable source of food and had another great advantage: its surplus could be stored for many months, even years. Corn provided the people of the southwest with the basis for a settled life.

The slow change from hunting and gathering to a life of farming in settled communities is a colourful and fascinating story, extending over

Corn grown by the Hopi people of Arizona still shows some of the many colours of the original varieties that were domesticated many thousands of years ago. *The most highly prized variety is blue corn, used for making special bread for ceremonies.*

A Hopi woman uses the traditional metate or grindstone to make flour from blue corn. *The flour will be made into piki, the thin sheets of bread eaten at weddings.*

many thousands of years, and involving the history of a number of well-known tribes. Here we have space only for a look at one of the most distinctive groups, the Anasazi, who occupied the territory around the so-called Four Corners, where the modern states of Utah, Arizona, Colorado and New Mexico meet.

The Anasazi found that the best places to live and grow crops in the desert were the canyons, where there was shelter from the winds and precious water flowed briefly after desert storms. To their staple, corn, they added the pumpkin-like squash, and beans, thus adding to the low protein of their diet. This existence, although still precarious, enabled the Anasazi to make quite significant changes in the way they lived. Their original pit houses—simple hollows in the ground roofed with logs and covered with mud—became deeper and more elaborate. Eventually, by about 1500 years ago, they began to live above ground in houses made of stone and adobe—mud bricks strengthened with straw. However, they retained the underground room, called a kiva, as a place for men's ceremonial activities.

Then, about 1000 years ago, dramatic developments took place among the Anasazi. In Chaco Canyon—a wide, shallow depression in the desert of New Mexico—they began to demonstrate a surprising capacity for public architecture and engineering. They erected buildings of a size and complexity that were quite unprecedented, and which were never excelled by any other society in North America. These building complexes, made of carefully laid stones, constituted what were virtually small towns, linked by a vast network of roadways. The Chacoan system eventually extended over more than 50 000 square kilometres. Even now, few modern Americans are aware of this highly developed form of urban architecture created by the 'redskins', who are still thought of in general as little groups of wanderers, sleeping in tepees.

The most impressive of the Anasazi ruins is Pueblo Bonito, a complex of buildings set against the north wall of Chaco Canyon. The centrepiece is a huge five-storey semicircular building that contained 650 rooms. It faced two plazas, each with its huge kiva, now open to the sky. Arranged around the periphery of the complex are the terraced rows of living quarters, their stone walls pierced by windows, and their rooms linked by doorways and passages. All the buildings are made of dressed sandstone and shale, superbly laid to make smooth walls and perfectly finished doorways and windows. More than a million pieces were quarried and dressed for the main building alone. Thousands of ponderosa pines were cut and carried for distances of up to fifty kilometres, to be used as beams. In addition, all the buildings were plastered, but this finish has now weathered away. The architects were well aware of seasonal variations in solar radiation; the buildings were sited so that the southward-facing doors and windows trapped the low winter sun for heating, but excluded the high summer sun.

A sustained campaign of archaeological excavation in Chaco Canyon over the past fifty years is only now beginning to disclose the true dimensions of this Anasazi 'state'. Pueblo Bonito, which could have housed

1000 people, is only one of more than 100 of these stone condominiums along the canyon floor. There must have been large and well-organised corn plantations to feed such concentrations of people. Without rivers to irrigate their fields, the Anasazi built extensive systems of diversion dams and ditches to collect and distribute the water which cascaded down the canyon walls during the rare thunderstorms.

For several hundred years these communities flourished, with a considerable trade in pottery, turquoise jewellery and stone implements. Then, quite suddenly, it all came to an end. The most likely explanation is a severe environmental change—perhaps a prolonged series of droughts, or erosion and loss of soil through the clearance of too much vegetation on the canyon floor. By about 1300, all the canyon sites were deserted.

The Anasazi moved back into the mountains of New Mexico and Colorado, and continued to live in pueblos. There are nineteen pueblos left today, but only one, Taos pueblo in New Mexico, near the headwaters of the Rio Grande, retains its original character. Its cluster of well-built houses are made of adobe bricks and timber beams. The largest, a multistorey dwelling house whose upper levels are reached by a series of ladders, is now the oldest continuously occupied building in North America.

The 1500 Tewa people who live in Taos pueblo, even as they try to obtain some benefit from the tourist trade, maintain the privacy of their kivas and their own lives. They were happy for us to film the handsome pueblo houses, many with their adobe bread ovens close by, but asked us not to film people. Behind the smooth brown adobe walls and closed pine doors they maintain a strong sense of their ancient traditions, not just clinging to a dying life-style but preserving the values and creativity that bound the Anasazi together.

The ruins of Pueblo Bonito, the great centre built by the Anasazi people in Chaco Canyon in New Mexico 1000 years ago. *Pueblo Bonito was one of a vast network of population centres that covered 50 000 square kilometres.*

Beams for the buildings in Pueblo Bonito were cut from ponderosa pines, and brought from more than fifty kilometres away. *The buildings were originally covered with plaster, but this has weathered away.*

The labyrinth of rooms and passageways in Pueblo Bonito illustrates the skills of the Anasazi stonemasons. *The largest building housed up to 1000 people.*

The main apartment house at Taos pueblo in New Mexico has been lived in for more than 700 years, and is the oldest continuously occupied building in North America. *The structure is made entirely of adobe—mud bricks reinforced with straw and dried in the sun.*

That creativity is alive and flourishing in the southwest today, through a revival of interest in the traditional Anasazi crafts of pottery, basket-making, and jewellery made from silver and turquoise. The Anasazi potters never knew the wheel, but made their bowls and pots by building up coils of clay, shaping their perfectly symmetrical vessels by hand and eye alone. The black pottery made by the same technique at the New Mexico pueblos of San Ildefonso and Santa Clara have achieved a high reputation and equally high value among collectors. The silver and turquoise pieces made at the Zuni pueblo and in Hopi communities in Arizona are justifiably famous for their designs and superb execution.[4]

In Arizona, too, the Hopi preserve the Anasazi tradition of making a particular kind of cornbread, called piki, for special occasions. This sacred bread is made from blue corn, one of the most prized of the many varieties of corn that are still grown in a dozen colours in the southwest. The kernels are ground into flour on a metate—the flat grinding stone—and mixed into a thin paste with a little ash from a desert plant. This is spread thinly on an oiled stone over a wood fire, and cooks in a few seconds into a crisp sheet, like tissue paper, which is folded into a pad. At Second Mesa, one of the main Hopi settlements, we watched a Hopi woman, formally dressed, and with the stylised Hopi butterfly hairbraids of the unmarried, tirelessly making and folding sheet after sheet with practised ease. She had an order of several hundred pieces for a wedding.

Other people who came later into the southwest, like the Navajo, are also keeping alive their traditions, such as rug-making. But the spirit of the Anasazi is all-pervasive—and nowhere more so than in the high canyons of Colorado, where they made their last stand in the thirteenth century. When everything collapsed in Chaco Canyon the Anasazi made a final attempt to keep their highly developed system going, by moving up into the high country and planting their corn on the mesa tops. Most of them lived on the green plateau of Mesa Verde, but for unknown reasons some descended into the giant canyon. There, beneath the vertical rock walls, they built an amazing series of mud-brick cliff dwellings, which for their sheer inaccessibility and air of mystery are perhaps the most remarkable structures in North America.

But in the end the unprecedented series of droughts, coupled with the onset of cold conditions which grew worse by the year, caused the abandonment of the Mesa Verde and the cliff dwellings, and the final scattering of the tens of thousands of Anasazi people. They could not have known it, of course, but what they experienced were the first signals of the onset of the 'little ice age', which was to grip the Northern Hemisphere for the next 400 to 500 years.[5]

The people of the southwest are well known, and they still have some of their land. But there are many other North American societies which did not survive—whole tribes and cultures that disappeared from the eastern seaboard, where their presence clashed with the interests of the Pilgrim Fathers and later arrivals from the Old World. Many are remembered now only by their tribal names: the Massachuset, Alabama, Delaware, Illinois, Iowa and Missouri.

There was, however, another whole group who have left no survivors at all. Even more curiously, their descendants were not even noticed in the great rush west after the birth of the United States. But in the areas they once occupied, the great river valleys of the Mississippi and Ohio, they have left astonishing reminders of their existence, in the structures which give these shadowy people their name: the mound-builders.

When the European settlers of the Atlantic seaboard of North America began the great westward movement away from the Atlantic states in the eighteenth and nineteenth centuries they began to come across unusual mounds in the great river valleys of the Mississippi and the Ohio. The mounds were covered with grass or bushes, and were not associated with any kind of building or other structure. Some of the mounds were quite small and looked like burial mounds. Others were in the shape of birds or animals, and some were enormous—up to thirty metres high. There were single mounds and in some places groups of 100 or more. Perhaps the most

The Hopi people still show strong physical affinities with their distant Asian ancestors. *The Hopis now live on their own land, and while friendly to visitors are increasingly seeking privacy to help them preserve their culture.*

The cliff dwellings of Mesa Verde in Colorado are among the most remarkable structures in the Americas. *These ruins were only discovered towards the end of the nineteenth century, by cowboys looking for lost cattle.*

intriguing was one in Ohio in the form of a giant serpent, its body coiling across a hilltop for more than 400 metres.

Settlers who were curious enough to dig into the mounds sometimes found human burials, together with stone tools and weapons, pottery items, even jewellery. The local tribes claimed that the mounds had been built by unknown people, long before their time. In any case, the settlers hardly thought the local people capable of organising the labour forces required to construct them. It was perhaps natural that all kinds of wild speculations were made about the identity of the mound-builders—Vikings, Romans, the ancient Greeks and Egyptians.

Today, after much archaeological research, we know that mound-building began more than 2500 years ago, with burial mounds. Excavations have unearthed amazing collections of artefacts buried as grave goods. These reflect not only the status of the persons buried, but also the extent of the vast trading networks that had developed over the eastern half of North America. There are finely polished stone axes and mace heads, delicate flint and obsidian arrow heads, ceremonial flint daggers, carved stone tablets, carved stone smoking pipes, pottery statues, animal and human figures cut from sheets of mica, carved shells from the Gulf of Mexico— even copper objects from the Great Lakes, far to the north, where people had discovered how to hammer native copper into various shapes.

The size of the mounds and the richness of the grave goods also reflected the change from hunting and gathering to a settled life and the growth of food, and the development of larger and more complex societies. With the introduction of improved strains of corn from Mexico around 1200 years ago, some astonishing mound-building societies appeared in the Mississippi and Ohio valleys. Many built mounds that served as platforms for temples and other public buildings. Out of complex social organisations grew ranked societies, with permanent positions of power which soon became hereditary. These were virtually city-states, with governments to run them.

The most awesome expression of this social development still stands beside the Mississippi at Cahokia in Illinois, just across the river from St Louis. Here, from about AD 600 onwards, there existed a large community of perhaps 20 000 people. They grew corn and other crops on the rich river flats, and with their labour built the most remarkable collection of mounds ever seen. At the centre was one gigantic mound with a base larger than the Great Pyramid in Egypt. Even today, after centuries of erosion, amateur excavation and attempted destruction, its summit still stands thirty metres above the surrounding landscape. Around it, in all directions, there are more than 100 other mounds. Many more have been destroyed and ploughed under to make way for farms. This city-state of Cahokia was the largest prehistoric city north of Mexico, with a trading network that extended from the Great Lakes to the Gulf of Mexico, from the Atlantic right out on to the Great Plains.

The construction of this mound complex represents a scale of social organisation and urban planning that many people, in America and elsewhere, still find difficult to credit to the original inhabitants of North America. Where they might have gone from here we will never know, because the arrival of the Europeans, and the diseases they brought with them, like smallpox, closed off all their options.

And yet archaeologists and anthropologists are also beginning to understand other changes, generated from within, which were beginning to affect those societies. The great advances that were being made—in population growth, in the development of systems of government, in architecture and town planning, in the building up of great trading networks—depended upon ever more intensive corn agriculture. But from some sites there is coming evidence of a price that was being paid for this progress.

The most striking evidence has been unearthed at Dickson Mounds in Illinois. The site gets its name from the Dickson family, whose members began to excavate the mound on their land in the 1920s and 1930s. They uncovered a remarkable number of human burials—men, women and children crowded together, some buried separately, others in groups. More than 250 human skeletons have been exposed, and left exactly as they were found, inside a modern museum which covers the site.

This find offered a rare opportunity for the anatomical study of a large population of people from the past, and the results have been quite unexpected and very revealing. Far from reflecting the healthy life of the active hunter, many of the skeletons showed bone deformations and other

Among the many different artefacts found in mounds are a beautifully flaked flint knife, and graceful shapes cut from sheets of mica. *These are in the Ohio State Museum in Columbus.*

signs of serious disease—tuberculosis and osteoarthritis, as well as malnutrition and anaemia during childhood. These latter effects were not severe enough to kill, but serious enough to leave telltale marks: unusual ridges on the teeth, shortened bones, holes in the roofs of the eye sockets. Many people had also suffered from the effects of overcrowding and bodily stress.

The Dickson Mounds skeletons are particularly interesting because they span the period from before corn agriculture arrived in the area until after it had become well established on the rich black-soil bottom lands. What they show is that the change from hunting and gathering to a sedentary farming existence certainly had its problems, and had a profound effect on the populations of the great river valleys—the mound-builders. Just as in our own society, there was clearly a hidden cost in the availability of a mass-produced diet, obtained without much physical effort.

Perhaps, therefore, it was poor health which began the breakdown of those great societies in the fifteenth century—a process which was fatally accelerated by the arrival of the Europeans with a whole new range of diseases (incubated in the crowded communities of Europe by precisely the same causes).

Far to the south, meanwhile, people in the other great American continent were also facing the breaching of their isolation, which had lasted for more than 30 000 years. They, too, had evolved some extraordinary societies, and had also developed a taste for great monuments.

Cahokia Mound in Illinois is the largest of the many thousands of mounds that were built in the Mississippi and Ohio valleys. *Although partly destroyed by amateur excavation and erosion, this mound is still thirty metres high, and is larger at the base than the Great Pyramid of Giza.*

The Dickson Mounds Museum in Illinois has been built over an early burial ground, dating back more than 1000 years, where the remains of more than 250 men, women and children have been uncovered. *Many of the skeletons show signs of dietary disorders, apparently due to the heavy reliance on a corn diet at that time.*

12.

Roads without Wheels

The continent of South America forms one of the major segments of the Pacific rim. From its northern edge, above the equator, down through the tropics into the chilly latitudes of Cape Horn—the distance from Siberia to Australia—South America displays extreme climatic and environmental variations. Behind its long and varied coastline it contains one of the world's longest and highest mountain chains, its largest expanse of rainforest, its greatest river system, and some of its hottest and driest deserts.

The people of South America are as culturally diverse as the habitats they have occupied for at least the past 30000 years. For much of that immensely long period they were virtually isolated from the rest of the world, so it is hardly surprising that the settlement of this vast continent remains one of the least known aspects of the peopling of the Pacific.

We have seen how the ice age people of Siberia, their physical characteristics acquired in a testing environment, passed through Beringia and into North America. Some of their descendants continued south down the west coast into Central America, then crossed the narrow isthmus into South America. These first arrivals in the new continent were of common stock, and would certainly have displayed the physical legacies of their Asian origins—dark straight hair, pale skin, and the mongoloid eye-fold. But as they fanned out across the new territory, to climb the high valleys of the Andes, plunge into the Amazonian rainforests and roam the vast open grasslands of the pampas, they began to develop societies and cultures that were different from any that had gone before.

Some groups kept to the Pacific coastline, heading south along the narrow plain between the ocean and the ramparts of the Andes. Eventually a few intrepid bands reached the archipelago of rugged islands that form the tip of the continent. At Cape Horn they stopped, because they could go no further. From the ancestral jungles of Southeast Asia, over the top of the world and down to the bottom again, that restless search for new horizons had brought people halfway round the globe. It was the end of

The 10000-year-old paintings in the Cave of the Hands of Patagonia tell part of the story of the long human occupation of South America. *The individuals who left their marks here did so by each placing a hand on the wall and blowing pigment from the mouth to make a stencil.*

221

humanity's longest journey.

The details and timing of the final leg of that epic migration remain obscure, because the rising sea level at the end of the ice age has covered the former coastline of North and South America, obliterating most habitation sites and other traces of human movement down the eastern side of the Pacific. We believe, however, that there is sufficient evidence from the few early coastal sites that do exist, and in the extraordinary range of surviving watercraft along the Central and South American coasts, to show that—as in other parts of the Pacific—the exploration and settlement of this continent was accomplished initially by sea, by people expanding quickly along the coast by raft, canoe and reed boat.

The last part of that colonisation, down to Cape Horn, must have been one of the most difficult. The ceaseless winds of the 'roaring forties' drive huge seas against the rocky coastline. Fierce currents sweep through the narrow passages between the islands. Glaciers spill off the snow-covered mountains and vast, swampy expanses of sphagnum moss block the valleys, making overland movement almost impossible. It is bitterly cold in winter, and even in summer there is always snow on the hills. The relentless winds prune the forests as if with shears, and in their gloomy interior there are no game animals or plant foods, except for one fungus which grows on tree trunks like exotic orange fruit. The swampy ground harbours no edible roots or burrowing animals. The streams flowing out of the peat bogs in the glacial valleys are so clear that there is little organic life in them, and therefore few fish or crustaceans. And yet in this demanding environment, which permits few mistakes, some people gained a foothold, because there is abundant food here—if you know where to look.

The cold and windswept Beagle Channel in Tierra del Fuego was the home of the most southerly permanent population that humans ever maintained. *The people here depended upon sea foods, and the entire mound in the foreground is made up of mussel shells from thousands of years of feasting.*

The Yahgan people of Tierra del Fuego survived in their severe environment for at least 12000 years, but succumbed rapidly to unfamiliar diseases when Europeans arrived. *The people wore skin cloaks in extreme conditions, but for much of the year went naked.*

When, in 1520, Magellan sailed through the strait which now bears his name, separating the mainland of South America from its tip, he named the large island to the south Tierra de los Fuegos—Land of Fires. (The name was later shortened to Tierra del Fuego.) The smoke plumes that Magellan saw rising everywhere were from the campfires of the local Ona people—one of the four tribes of the most southerly human population in the world. In that forbidding climate the people kept fires forever burning in their shelters.

The Ona were nomadic, wandering in small family groups on the trail of guanaco, on which they were dependent. Guanaco are graceful animals of the camel family, with long, soft fur, and are widely distributed in South America. For the Ona, vegetable foods were scarce, and they never used canoes, harpoons or fish spears, so that apart from an occasional otter, fox, swan or duck the guanaco was their only food. The Ona were excellent woodsmen, and stalked their shy quarry with bows and arrows. The guanaco skins were made into cloaks and moccasins. Cleaned of hair and made waterproof with mud, the hides were used to make windbreaks, which were the only form of shelter the Ona had.

Further south were three other tribes, the Yahgan, Haush and Alacaluf. These southerly people had a close relationship with the sea, and depended on the abundant marine resources which flourished in the cold but rich waters around the islands of the archipelago. They were accustomed to a much hardier existence than the Ona, and although they sometimes wore fur cloaks they went naked for most of the year. The Haush were a small group, confined to the eastern tip of Tierra del Fuego, and the Alacaluf lived among the remote islands on the west or Pacific side. The main tribe were the Yahgan, who lived along the southern edge of Tierra del Fuego on the beaches of the Beagle Channel, and on the smaller islands further south, right to Cape Horn. They never ventured inland, to the north, for fear of the Ona.

The Yahgan were venturesome marine hunters, in waters that were always windy and often very rough. They made their canoes from thick sheets of bark, and kept a small fire burning in a bowl of rocks in the centre of it while they travelled. Entire families joined in the hunt. The women rowed the canoe, with the children in the centre, while the men crouched in the bow with their spears and harpoons, as they pursued seals, sea lions and steamer ducks around the rocky islands, and otters in the bays and estuaries. Occasionally they found a whale or elephant seal, for the Antarctic Peninsula was not far away.

The Beagle Channel was the main axis of the Yahgan world, and to follow the shores of the Channel today is like walking along the west coast of Tasmania. There is midden after midden of mussel shells and fish bones, some with hollows in the top where the Yahgan had built their domed huts, roofed with sticks, just as the Tasmanians had done. The Yahgan women had similar foraging skills to the Tasmanians, diving for mussels, whelks and octopuses, trapping huge king crabs, and collecting edible seaweed. Their simple tools of stone, bone and shell were not much more elaborate than the basic Tasmanian tool kit, and they also wove well-made baskets. The simplicity of the Yahgan economy was relieved by their quite elaborate ceremonial life. They painted their bodies with vivid stripes, and some of the dances and ceremonies lasted for months.

The earliest known archaeological sites in Tierra del Fuego are around 13 000 years old, so it is clear that these hardy southern people continued their hunting life, and their isolation, for a very long time. That isolation was maintained for a time even after the Spaniards arrived in South America in the sixteenth century, because the arid deserts of Patagonia to their immediate north acted as a barrier to European settlement. But in the end the isolation of the Fuegans proved their undoing, for when the Europeans did arrive the people had no defences against their guns or, more significantly, their diseases. The Fuegans and their fragile cultures— the most southerly permanent outpost of the human family—were wiped out within a few decades. (In 1884 half the Yahgan population of 1000 died of measles in three months, immediately following the establishment of an Argentine settlement at Ushuaia, on the Beagle Channel.)

The people at the southern tip of South America were remarkable enough in their adaptations, but they were only a tiny segment of the broad spectrum of human cultures and societies that evolved across this vast land mass. A few of these early cultures remain, although today they are fast disappearing—generally in the face of economic development.

Once there were fishing communities all round the coast, but one of the very few to maintain anything like their original way of life are the coastal Peruvians, who still make and use their reed boats. They now live in modern houses in a suburb of a large city, but they still go out and cut bundles of reeds from swamps near the beach and tie them into the one-man boats with upturned prows that they call 'caballitos'—little horses. The reason for this is clear when you watch the fishermen go out in them, kneeling in a shallow depression cut near the stern and paddling hard. The boats rear up as they meet the incoming surf, the fishermen riding them like horsemen. The great buoyancy of the reeds carries the caballitos safely through quite large waves. The fishermen use them to catch crabs just outside the surf line by lowering baited baskets to the bottom.

The boat-building tradition is obviously very ancient in South America. On Lake Titicaca, on the border of Peru and Bolivia, reed boats are also still in use. In the National Museum of Chile in Santiago there is a remarkable boat once used by seal hunters. Cactus thorns were used like pins to join several seal skins into two sausage-shaped chambers, pointed at one end. These were inflated through a tube, and linked by a wooden platform, to form a stable and easily paddled craft—the prehistoric forerunner of the modern Zodiac inflatable boat. For thousands of years the South American coast was obviously travelled by all kinds of water-craft, including dugouts, caballitos, rafts of animal skins and bamboo, and those great sailing rafts of balsa logs, made famous by Thor Heyerdahl and *Kon-Tiki*.

While this kind of evidence supports the idea of an initial human

movement down the Pacific coast of South America, there is little or nothing to suggest when the first people entered the Amazonian basin—with the exception of one startling archaeological site in eastern Brazil, which has challenged the conventional view of American prehistory.

The site is actually a series of excavations that have been carried out over the past ten years by a joint French-Brazilian team involving thirty-five specialists in archaeology, geology, ecology and related subjects. They have been investigating a series of rock shelters along a line of sandstone cliffs containing rock paintings and engravings of human figures and animals, and human habitation sites, as shown by the presence of fire-places, stone tools, food storage pits and pottery. So far 260 archaeological sites have been found, most of them containing art.

What has caused surprise—and scepticism—among American archae-ologists and prehistorians are the dates being established at the main site, a large rock shelter at a place called Pedra Furada, containing more than 1000 painted figures on its back wall. There, in an excavation at the foot of a large panel of rock paintings, the archaeologists found a succession of human occupations, indicated by fireplaces and tools such as knives, choppers and scrapers, in layers going down more than five metres. Charcoal samples from hearths at different levels have yielded a series of consistent carbon-14 dates that range from 32000 to 17000 years ago. In the surrounding area the archaeologists have found more stone artefacts and fireplaces, in association with the bones of extinct ice age animals, such as giant ground sloths.[1]

It will be appreciated that such dates are difficult to reconcile with the widely held view, based on dated sites in North America, that the first people did not enter the Americas until shortly before 12000 years ago, when the northern ice sheets retreated and opened up corridors from Beringia. Such early dates are, however, not inconsistent with a rapid colonisation of the Americas by sea along the coastlines, beginning before 40000 years ago. It can be seen from the map that the easiest way to reach the Pedra Furada area from the Isthmus of Panama would be along the northeast coast of South America, and up the Rio Parnaiba, which rises not far from the site.

The other interesting aspect of the Brazilian sites is the age of the rock paintings. Fragments of painted rock from the back wall of the shelter had apparently fallen down during some of the earliest occupation phases, for they were found between layers 27000 to 32000 years old. This makes the paintings the oldest in either South or North America. There are other paintings in South America which, although not as old, are particularly evocative of past times. These are high up on the side of a spectacular canyon in the wilds of Patagonia, in a huge rock overhang known as the Cave of the Hands.

Here, extending for nearly 100 metres along the cliff above a ledge, is an astonishing gallery of human hands, stencilled on the rock surface in red, black, white and yellow. Most were made by people spraying a mouthful of ochre pigment over their hands; some were made directly, by pressing a

The reed boats used by coastal fishermen in Peru are one of the oldest forms of watercraft in the Americas. *Similar boats are used on Lake Titicaca in the Andes, and craft of similar shape, although made of rolls of bark, were once used in Tasmania.*

paint-covered hand on the wall. There are also paintings of animals such as the guanaco and armadillo, and the giant condor, which still sweeps through the canyon with its shrill cry.

What imperative brought all these men, women and children to this remote rock ledge, high above the canyon floor, we can only guess at. Hand stencils are a universal art form, and there are galleries like this—but rarely on such a scale—in Australia, North America, Africa and Europe. They are a means of personal identification, recording the presence of the artist as surely as a fingerprint or a signature. These paintings in the Cave of the Hands have been dated to about 10 000 years ago, not long after the end of the ice age—and yet, because of their freshness, preserved by the dryness of the desert air, they have an eerie immediacy, as if the owners of the hands had just passed by, and their voices had just faded round the bend of the canyon. Yet they also speak of a time and a way of life that has ended, when small bands of hunters roamed the plains nearby in search of guanacos, rheas and armadillos, and giant sloths and pumas haunted the caves and defiles below the galleries.

As in other areas of the Pacific basin, towards the end of the ice ages, the South Americans began to manipulate plants and animals to their advantage—to initiate the process of domestication. The diversity of their vast continent, and its huge range of plants, offered many kinds of fruits and vegetables for experimentation. Exactly when this began is of course still unknown, but from the results we can be sure that it began here at least as early as anywhere else.

In 1976 Chilean and American archaeologists made a discovery in southern Chile which indicates the kind of gradual moves towards domestication that must have been taking place in many areas. At a place called Monte Verde they found an ancient campsite beside a stream which runs through the coastal forest. Carbon dating indicated that people had certainly lived there 13 000 years ago, and had possibly even used the site as early as 30 000 years ago.[2]

The Monte Verde site was unusual in that it was a waterlogged peat bog, and the lack of oxygen had preserved artefacts of organic materials which normally rot away and disappear. The crucial find from the 13 000-year-old level was of cut logs and branches laid out in squares—clearly the foundations and frames of huts or houses, which would have been roofed with animal skins. There were also stone choppers and knives, wooden spear points and digging sticks, and bones of animals which had been eaten there. These included the extinct mastodon, and a camel related to the guanaco and its domesticated forms, the llama and alpaca. The bones of some of the animals, and some elephant tusks, had been fashioned into tools of various kinds. Obviously, what had been found was a site of a village or settlement, occupied by people in a transition stage from hunting and gathering to a settled life, well before the end of the ice ages.

Among the requirements of such a settling-down process are reliable, accessible, year-round supplies of food, and this is what the country around Monte Verde offered. The people were obviously still hunting,

The southern end of the Andes, between Argentina and Chile, contains some of the world's largest glaciers. This one periodically pushes across Lake Argentino to the near shore, forming a dam. *Eventually the enormous pressure of water causes it to burst, with spectacular results.*

because they had spear points of various kinds. As well, they had bolas — smooth, rounded stones with grooves around them, which were tied to string, whirled round the hunter's head, and then flung at running animals to entangle their legs. (The bola is still used in this way in parts of Chile and Argentina.) They also caught crustaceans in the stream and nearby lakes, and fish and shellfish in the sea, which was not far away.

But the mixed forest of deciduous and evergreen trees also offered a variety of plant foods in all seasons — fresh shoots in spring, berries in summer, nuts and seeds in autumn and bulbs in winter. All these were found at Monte Verde, but the most interesting remains were some small dark specks that were unmistakably fragments of potato. These were from wild species, but they were very similar to the particular wild form which was later domesticated in the Andes.

At Monte Verde, therefore, we can see the beginnings of several crucial changes that would transform human activities on this continent. People were beginning to construct permanent housing (and so far no earlier housing has been found in North or South America), and some of them probably stayed at the site all year round. And although they were still hunting they also brought back two wild foods—the camel and the potato—which were about to be domesticated. It is the earliest glimpse of village life that we have from anywhere in the Americas.

Domestication, when it did come in South America, was to transform the lives of every group it touched. To the east of the Andes, in and around the Amazonian basin, the food plants brought into cultivation included peanuts, pineapples, passionfruit, arrowroot and, most importantly, the cassava, which produces a large, starchy root with exceptionally high calorific value. This is known as manioc in South America (and when processed, as tapioca), and it helped the hunter gatherers of the Amazon make the transition to cultivation, because it grows well on the rainforest floor without much attention.

It was, however, in the high Andes of northern Peru, and the coastal plains close by, that the crucial domestication of plants and animals was to take place. As a consequence, it was these areas that saw the rise of the most complex societies in the whole of South America.

There is a good reason why the lowlands of Peru, beginning about 6000 years ago, were an early centre of domestication of such plants as cotton, lima beans and various kinds of squash. The cold coastal currents from the Antarctic brought sufficient nutrients to support huge populations of fish, seals, dolphins and birds, and these marine resources were an insurance for the coastal people against poor growing seasons or even wholesale crop failures.

It was in the highlands of Peru, however, that the most important domestication took place, of both plants and animals. The crucial food plant brought into cultivation, which was destined to become one of the eight staple foods in the world, was the potato. (The others are all grains: rice, wheat, maize, millet, oats, rye and barley.) All the members of the large family of wild tubers from which the cultivated potato is descended have varying amounts of toxic alkaloids in them, and the process of domestication, which is thought to have begun somewhere around 7000 years ago, must have begun with the selection of wild tubers that were less poisonous. And that diversity persists, even today. In Peruvian markets you can see potatoes of a dozen shapes and colours—shades of red, yellow, brown and grey. The traditional storage method now, as it was thousands of years ago, is to trample the potatoes and expose them to the dry mountain air; in this way they are effectively dehydrated, and can be kept for long periods.

Side by side with the domestication of the potato the Andean people first tamed the wild guanaco, and then produced from it, by selective breeding, two extremely useful animals, the llama and the alpaca. The llama is the larger of the two, and its erect, dignified, even haughty demeanour is as unmistakable a trademark of Peru as the kangaroo is of Australia. Its fur is thick but coarse, but its meat is palatable. Its most important role in the Andes, however, has been as a pack animal. For thousands of years the llama has carried loads through the steep mountain valleys where no other

The llama, a domesticated member of the camel family, was produced in Peru thousands of years ago by selective breeding from the wild guanaco. *The llama's meat is useful, but its most important role in South America has been as a pack animal.*

Guineapigs were domesticated in the Andes for food, and most houses in mountain villages have some living in the walls or under the floor. *Even today, guineapigs provide most of the animal protein that Peruvian villagers get.*

transport was known.

The alpaca is smaller, but its coat is much finer and yields superb wool for spinning and weaving. The Andean women have made clothes and blankets from alpaca wool since prehistoric times, and today it makes some of the most expensive overcoats to be found in the shops of North America and Europe.

One other animal that was domesticated long ago in the Andes is a small rodent that now lives with people in Peruvian villages, in tunnels in the walls and under the floors of the stone and mud-brick houses. It is plump and furry, and breeds very quickly when fed on scraps from the table, or fresh grass. When a few months old this animal makes excellent food, and is one of the major sources of animal protein for Andean people. It is usually eaten roasted, and can often be seen on food stalls in markets, cooked ready for use. The Peruvians have their own names for this handy item of food, but elsewhere in the world it is known as the guineapig.

From the foundations of a large array of cultivated food plants, an abundance of sea foods, and a number of useful domesticated animals (as well as assets which would come into their own later: immense resources of copper and gold) the South Americans created along the Pacific coast a brilliant series of societies that produced startling bursts of creativity in pottery, fabrics, architecture and metalwork.

One of the most sophisticated of those early cultures, the Moche, spanned the period from the beginning of the Christian era until about AD 600. The lives of the Moche are uncannily revealed to us because they happened to be master potters, and in making pots to accompany the burial of prominent people they created astonishing likenesses of actual

We can get a vivid picture of the Moche culture, which ruled northern Peru from the beginning of the Christian era until about AD 600, from the thousands of pots found in burials, depicting real people and scenes of everyday life. *These pots show three prominent men, a woman washing her hair, and a prisoner receiving special punishment as a bird pecks out his eyes.*

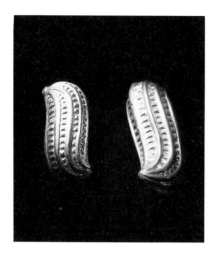

This funeral mask and pair of peanut ornaments, found recently in northern Peru are exquisite examples of Moche gold work. *The looting of treasure from burials has become so prevalent that finds of unlooted tombs are becoming increasingly rare.*

people, animals, fruits and vegetables, and recorded many human activities of those times. There are Moche pots that are portraits of kings, warriors, priests, as well as ordinary individuals—each one different and particular. There are pet dogs, rabbits, ducks, lizards and frogs. There is a woman washing her hair, a midwife assisting a birth, a doctor examining a patient, even a bird pecking out the eyes of a criminal—a special form of punishment. And there are all the foods—potatoes, corn, chillis, peanuts, pineapples and seals, crabs, fish and lobsters.

The Moche were also the first of a succession of highly skilled metalworkers in South America. Many of the finest early works in gold and platinum were buried in tombs, and have been lost over the centuries to looters. But the recent discovery of a virtually intact tomb of a high Moche noble, buried in the third century AD, has yielded the most spectacular hoard of gold and turquoise treasures ever found in any burial in all the Americas.

Another discovery in northern Peru, not far from the Moche tomb, has solved one of the most tantalising mysteries of the history of metalworking in the Americas. There is a long history of working gold and platinum in South America, gold and silver in Central America, and native copper around the Great Lakes in North America. It is also known that the smelting of copper from its ores was discovered in South America perhaps 2000 years ago. But although hundreds of copper and bronze artefacts have been unearthed, archaeometallurgists have never been able to find

In a re-creation of ancient copper smelting techniques, men with cane blowpipes raise the temperature in a 1000-year-old furnace at Batan Grande in northern Peru. *The area contains hundreds of bowl furnaces, and is the first metal smelting site to be found anywhere in the Americas.*

Small globules of pure copper in the furnace slag at the end of the smelt at Batan Grande. *Skilled and experienced smeltermen could obtain up to 500 grams of copper from each smelt in these furnaces.*

any evidence of copper smelting.

Then, in 1979, in an arid valley near the village of Batan Grande, a party of American and Peruvian archaeologists came across the evidence they had been seeking for years. On a hillside they found first one, then several, then literally hundreds of small bowl-shaped clay furnaces, about as big as a litre jug, set in the ground. The clay linings were blackened from firing, and some furnaces contained fragments of charcoal and pieces of copper slag. Nearby were large flat grinding stones (batanes) and rocking stones (chungas), which had been used to crush the copper ore ready for smelting (hence the name of the local village, Batan Grande). Carbon dating showed that the furnaces had been used from around AD 800 until about AD 1200 or 1300. It was the first smelting site to be discovered in the New World.

Only one part of the jigsaw was missing. Since it was known that the South Americans had never had bellows, how had they achieved the required temperature of more than 1000 degrees Celsius in their furnaces? Further search around the site disclosed a number of clay nozzles, about three centimetres in diameter. Nearby swamps yielded large reeds whose tubular stems were about the same size. Dr Izumi Shimada of Harvard University, the leader of the team, realised that the early smeltermen must have used teams of people with blowpipes to provide the furnace draft. The clay nozzles or tuyeres pushed into the end of the blowpipes would prevent them from catching fire.

In 1986 and again in 1987 Dr Shimada organised an experimental smelt in one of the thousand-year-old furnaces at Batan Grande. He used copper oxide (malachite) from deposits nearby, charcoal made from local trees, reed blowpipes with clay tuyeres, and a team of men from the village to blow. We filmed the second of these experiments.

As in ancient times, the furnace was filled with charcoal and copper oxide, with some iron ore to act as a flux, and the smelt began, with three or four men blowing simultaneously through their blowpipes. The glowing charcoal produced carbon dioxide gas, which created a 'reducing' atmosphere in the furnace, drawing out the copper from its ore in tiny globules, which remained trapped in the mass of charcoal and slag. After four hours of continuous blowing, and the addition of extra copper ore and charcoal, the smelt ended. When the lump of slag in the bottom of the furnace cooled, Dr Shimada pulverised it with a chunga on one of the batanes to extract the globules or prills of pure copper. On his first smelt he had obtained a good handful of copper; on this occasion only a few prills. But it was enough to show that the Peruvians had indeed made the quite independent discovery of how to smelt copper, and that South America experienced its own bronze age until the fifteenth century, and the arrival of the Europeans.

It is ironical that the Spaniards got to South America just as the greatest and most extensive empire that the southern continent was to produce had reached its dazzling climax. That empire—the only major society to develop entirely in the highlands of the Andes—was ruled by the sun god, the Inca.

For a people who were to create, quite suddenly, such a vast and powerful empire, the Inca had extremely humble origins. They began as a small chiefdom in the Cuzco area, about AD 1200. The word Inca originally referred to their ruler, who was considered to be the incarnation of the sun. The early Inca state seems to have been like many others in the Andes, made up of peasant farming families growing potatoes and raising guineapigs, perhaps keeping a few alpacas and trading their woven woollen blankets with nearby communities. But then, by some human alchemy or combination of circumstances, the Inca were transformed into an energetic, powerful, expanding society.

They evolved a rigidly hierarchical system of government, and as their numbers and territory grew they established provincial capitals. Over the next 300 years huge areas of territory were added to the Inca domain. As land was acquired its population was reorganised according to Inca principles, with governors for every 10 000 people. Taxation was largely paid in labour, thereby expanding food production and construction programs. By the beginning of the sixteenth century the Inca dominated an area that stretched for nearly 2500 kilometres along the spine of the Andes and the Pacific coast. To the north it reached what is now Colombia. To the south it included Ecuador, Peru, Bolivia, and large portions of Chile and Argentina. Altogether, the Inca empire extended from the equator to latitude thirty-five degrees South. It embraced freezing altiplanos or high plateaus, deep river gorges, steamy jungles in the Amazon basin, and the scorching deserts of the Atacama.

One of the greatest monuments to the Inca empire is Machu Picchu, built high among the peaks of the Andes. *Machu Picchu was never discovered by the Spanish conquistadors, but with the collapse of the Inca empire it was abandoned, and not rediscovered until 1912.*

One of the great Inca skills was the way they were able, using only stone pounders, to shape and fit the huge granite blocks in their walls and buildings. *It is instructive to try to work out which of these blocks was made first.*

To link the elements of this far-flung domain the Inca built thousands of kilometres of roads, many with paved stone surfaces. They crossed ravines on suspension bridges, rivers on low stone causeways, and lakes and marshes on pontoons. But these roads saw no wheels, because although the Inca invented wheeled toys for their children they never made full-scale wheeled vehicles—perhaps because they lacked a suitable animal to pull them. They had, however, a useful transport force in their herds of llamas. These animals could carry loads of up to forty-five kilograms, and long trains of them docilely carried the commercial traffic of the Inca empire. They shared the roads with an amazing amount of human traffic, most of it on official Inca business: messengers and couriers, armies on the march, whole populations of conquered civilians being relocated.

Among their many technical skills the Inca engineers developed an astounding facility for building in stone. They devised techniques of cutting and shaping huge blocks of granite weighing up to 100 tonnes so accurately that when they were fitted together in a wall or building a knife-blade could not be slid between them. The Inca had few bronze tools (and no iron at all), and are believed to have used stone pounders for all their stone work, finishing the blocks by hand-grinding with wet sand. The enormous blocks must have been hauled up and lowered down the steep valleys by ropes, perhaps on ramps of earth. But despite these handicaps the Inca built an extraordinary range of palaces, temples, walls, terraces and public buildings across their empire.

The other great obsession of the Incas, and one which eventually brought about their downfall, was gold. In the incredibly rich alluvial deposits in the Amazonian rivers they had access to virtually limitless quantities of gold, and they made from it a treasure-house of gold objects such as the world had never seen. They did not regard gold as wealth or a negotiable asset, because they never used it as currency. They admired its glowing colour, its smooth, workable surface, and its incomparable lustre. It was, like their ruler, the incarnation of the sun, and they swamped him

In the absence of any kind of draught animal to pull a plough, the Andean people devised the foot plough to cultivate their fields. *This ancient device is still widely used in Peru.*

In the Valley of the Incas in Peru life has changed little in the past 1000 years. *These women live in Ollantaytambo, a village that has retained its Inca character.*

in gold ornaments of every conceivable kind. And it was a similar obsession with gold, although of a very different kind, which brought the Spanish conquistadors to Cuzco in 1531. Soon after, the whole Inca world came crashing down.

In many parts of Peru there are still vivid reminders of the Inca—the ruined citadel of Machu Picchu, lost in the mists until this century; lonely, rock-strewn stretches of the great road system; the massive walls of the fortress of Sacsahuaman above Cuzco; the dress and customs of the people in ancient villages like Ollantaytambo. In the Valley of the Incas that carries the Urabamba River life goes on very much as it did under the Inca. Men use the ancient foot plough to turn over their potato fields. Women sit in the village squares weaving alpaca wool on their backstrap looms. High up on the valley walls tiny figures work on the narrow terraces. In the markets the sellers display vegetables never seen outside the Andes. There are bundles of coca leaves, too, which in the Andes are chewed to ward off cold and hunger, and as a sacred comfort against evil spirits. Few here would understand or forgive the way the rest of the world now uses the coca plant's derivative, cocaine.

But to us the most inspiring reminder of the lost Inca way of life is to be found far from Cuzco, in a remote and virtually uninhabited part of the southern highlands of Peru, accessible only across passes at 5000 metres. Here, the gigantic Colcha Valley cuts a gash in the earth deeper and wider than the Grand Canyon of North America. And here, 500 years ago, the Inca rulers decreed that the canyon walls be terraced and irrigated to feed the growing populations being brought under Inca domination.

The engineers of those times performed an astonishing feat of construction and water management. Streams in the high valleys were trapped and led along stone conduits below the canyon rim, then fed by gravity to the fields of crops on terraces that formed giant staircases down to the

In Peru the fine wool from the alpaca is spun each summer to make warm clothes for the winter. *The sheep-sized alpaca and its larger relative, the llama, are both members of the camel family.*

valley floor. It was one of the most ambitious and brilliantly executed systems of food production ever undertaken. But when the Spaniards came the Inca abandoned the terraced fields and filled the irrigation channels with rocks. The Colcha Valley was returned to nature.

But in recent years, with the growth of the Peruvian population, the government has faced a seemingly insoluble problem in feeding its people. Attempts to irrigate more of the arid coastal plain, in order to open up new farmlands, have proved hugely expensive. Remarkably, the government has found that for a fraction of the cost per hectare the ancient Inca irrigation channels and terraces in the Colcha Valley can be brought back into production.

Archaeologists have been excavating the Inca constructions along the canyon sides, discovering how the system worked. Now local villagers and farmers are cooperating in rebuilding the old terraces and restoring the fields, preparing for the water that will soon be cascading once again down

The terracing of the Colcha Valley in southern Peru by the Inca was one of the greatest works of engineering in the Pacific basin. *Some of the terraces are being brought back into production to supply food for Peru's growing population.*

the 500-year-old channels. Some sections have already been restored, and green patches of crops are appearing once again on the terraces and on the valley floor. Like the water, the spirit of the Inca seems to be flowing through the Colcha Valley once again.

Around the Pacific basin we have seen many examples of cultural continuity, linking the past with the present. But few are as direct or as productive as they are in the Colcha Valley, where the technologies of today are no more efficient than those that were developed more than 500 years ago. In its conception and execution, in the sheer physical scale of it, the Inca works in the Colcha Valley are one of the greatest human achievements anywhere in the Pacific basin.

13.

The Feathered Serpent

The southern end of North America, where the continent narrows down into the Isthmus of Panama, was the highway south for the first Americans. For many thousands of years after that initial migration into South America there was a constant movement of people, products and ideas back and forth through this geographical bottleneck—an exchange of inventions, artefacts, and the new foods and animals as they were domesticated.

Not surprisingly—perhaps even inevitably—there developed in central or Mesoamerica, and particularly in Mexico, a series of the greatest civilisations that the Americas were to produce. What set them apart from other cultures around the Pacific was their rapid rise, flowering and decline, one after the other, in a relatively small area. In China over that same period, from about 1000 BC to the sixteenth century AD, there was a single civilisation. In Mexico, however, a number of closely related people produced a sequence of states and empires whose archaeological remains are so spectacularly different from one another that we tend to think of them as unrelated. But what was to become a dazzling succession of cultures had very humble beginnings, in the valleys of central Mexico.

By about 10 000 years ago, soon after the end of the ice ages, the herds of large grazing animals—mammoths, bison and horses—had disappeared from much of Mesoamerica, as they had from North America, through a combination of human hunting and a gradual drying out of much of the country. With their hunting options reduced to smaller animals like deer and rabbits, the human populations began to broaden their subsistence base by bringing a number of favoured plants and animals under control. The best evidence of this process, and what it was to lead to, has been found in the Tehuacan Valley of central Mexico, about 100 kilometres southeast of Mexico City.

Today the Tehuacan Valley is a dusty green oasis of farms and patches of cottonwood trees, ringed by the arid, cactus-dotted slopes of the Sierra Madre, scene of many a Hollywood western. At the end of the ice age it was

The series of civilisations which evolved in central America or Mesoamerica formed a dazzling cultural succession, in which one powerful state after another rose and fell with dramatic finality. *The first advanced society was that of the Olmec, whose rulers were portrayed in massive stone heads that weigh up to twenty tonnes.*

more moist and more verdant, and there were many different edible plants available for domestication by the wandering hunter gatherers in the region. Archaeological finds in caves high up on the slopes of the Sierra Madre show that the first plants to be cultivated were members of the family of cucurbits, which include gourds, squashes and pumpkins.

The fruits of the wild species of these plants are large and often brightly coloured. They obviously attracted the attention of foraging people, and there is evidence that they have been a part of human diet in Mesoamerica for at least 10 000 years. At first, only the seeds were eaten, because the rinds were hard and tough and the flesh was bitter. In sampling the wild fruits for seeds, however, those early Mexicans would no doubt have found occasional mutants whose flesh was less bitter, and so would have begun the long process of selection which produced domesticated varieties. The earliest to be grown regularly was the bottle gourd, which was useful for more than just food. The shells were (and still are) used as containers for liquids and food, for storage, and for making eating utensils and musical instruments.

Some time later the people in these valleys began to plant beans, chilli peppers and avocados. The wild avocado had small fruit with large seeds, but human selection gradually produced a fruit very similar to the one known today all round the world.[1] From the bones found in habitation sites it is clear that the Mexicans had begun to tame wild dogs and turkeys, also for food. And about 7000 years ago another domesticated plant appears in the archaeological record in the Tehuacan Valley—the most important American contribution to the human diet, not just in Mesoamerica but all over the world. This plant was maize or corn (*Zea mays*).

There is still considerable debate about how corn was first domesticated,

Mexico was one of the earliest centres of crop domestication in the Americas, where corn, beans, tomatoes, squash, avocados and cacao were brought under cultivation. *These crops are growing over a mound at the Olmec site of Tres Zapotes, where farming was practised more than 3000 years ago.*

because it has no clearly identifiable wild ancestor. The very similar grain-bearing annual teosinte, which grows wild in Mesoamerica, is a close relative, but there are fundamental botanical differences. The seeds of teosinte are very loosely enclosed by a few husks, and therefore scatter naturally when ripe, while those of corn are tightly enclosed and remain on the spike. Corn is therefore totally dependent upon human agency for its propagation. One suggestion for its origin is that corn, teosinte and another wild grain, tripsacum, are descendants of a common ancestor, which has now completely disappeared.

Whatever the explanation, it is known that corn—although its cobs at that time were not much bigger than a man's thumb—transformed life in Mesoamerica. As a crop it was relatively easy to grow, and people began to settle down in open villages beside the corn patches, marked in so many archaeological sites by the pairs of manos and metates—grinders and grindstones. (By the time the Spaniards arrived in the Americas corn agriculture had spread north as far as Canada and south, across the isthmus, as far as Chile.)

Pottery is a marker of village life, because it is too fragile to be carried constantly from place to place, and it began to appear about 4000 years ago. By 3000 years ago people were living a settled village life throughout Mexico and areas to the south. They had a variety of stone tools for processing food, and had developed basket-making. By then powerful groups of people had emerged, and were starting up trade between different areas of the highlands and the swampy coastal forests and plains on the Gulf of Mexico. The highland villagers were growing cotton and making sisal cordage and the coastal people were growing cassava and harvesting the sea for food and shells, and both kinds of goods were being traded. By this time the Mesoamerican diet had been considerably diversified by adding yams, tomatoes, papayas, cashew nuts and guavas. Some people were also cultivating the cacao tree, and having discovered how to process the seeds from its pulpy fruits into chocolate it is not difficult to see why they regarded this plant as having come to them from paradise.

It is one of the mysteries of the human story in Mesoamerica that the first advanced society arose not in the higher valleys, where the earliest evidence of community life has been found, but in the steamy, lowland jungles along the coast of the Gulf of Mexico. One factor may have been the higher annual rainfall, which produced regular flooding of low-lying areas. The receding waters left a layer of rich mud, and this Nile-like situation would have permitted more intense cultivation of crops and supported a larger population. However, the conditions and circumstances which brought this first Mesoamerican high culture into existence are still not properly understood. All we know is that it happened about 3000 years ago, and that the evidence for it is extremely solid—monumental, in fact.

The Olmec civilisation was first recognised principally by its distinctive sculptural style, which combined powerful, massive form with simplicity of treatment. Isolated finds in the lowland jungles of a gigantic stone head and some small jade figures with half-human, half-feline features had intrigued archaeologists in the early part of this century, because they were clearly different from other Mesoamerican cultures. But there was no way of even relating them to one another, because the jungle had seemingly swallowed up all traces of the people who had made the carvings. Then, in 1925, two American archaeologists began exploring southern Mexico in search of 'a place of idols and temples' along a river, briefly mentioned in a Spanish journal of 1518.

The two men were picking their way through swamps and dense undergrowth prowled by jaguars and infested with snakes when they suddenly came upon an astonishing scene on a small island in a river. Hung with creepers and green with moss, their details crumbling but still strikingly obvious, there were extraordinary stone structures—massive altar-like stones, some with figures of rulers carved on them; a tomb of huge basalt columns stuck in the ground like tree trunks, and containing human burials decked with jade ornaments; stelae or pillars carved with figures of powerful leaders and jaguars, serpents and alligators. There was also a great bell-shaped boulder which, the archaeologists wrote later, 'puzzled us very much, but after a little digging to our amazement we saw that what we had in front of us was the upper part of a colossal head'.

The huge head was only one of four found at the site, called La Venta. The largest is 2.3 metres high and weighs more than twenty tonnes. The faces, with thickened lips and flat noses, are reminiscent of some Southeast Asian populations. Twenty-one of these colossal heads have now been found along the Gulf coast, and it is now virtually certain that they are portraits of Olmec rulers. Each wears a slightly different helmet, not unlike those worn by American football players. Some authorities believe that the helmets may have been worn in a sacred ball game, which we know from clay figurines was played by the Olmec. (The name Olmec translates as 'rubber people', and may have referred to the ball, made of natural latex.)

The site at La Venta—including a huge cone-shaped mound thirty metres high—has now been destroyed by oil exploration, but all the Olmec objects from the site have been moved to a park in the nearby city of Villahermosa. There, these masterpieces of sculpture have been set up in a jungle atmosphere, complete with monkeys, deer and other animals. Other finds recovered during excavation at the La Venta site, and now on display, include three striking mosaic pavements showing stylised jaguar heads. Since 1925 many Olmec sites have been located, and it is now clear that these people developed the first Mesoamerican civilisation. It was the first flowering of the key aesthetic and philosophical elements that were to be repeated again and again in Mexico by later cultures.

The Olmec culture began just over 3000 years ago and lasted perhaps until the beginning of the Christian era. It was based on corn, which in the wet coastal belt could be grown all year round. The figures carved on the stone stelae indicate the development of a highly stratified society, with powerful rulers, aided by priests, controlling the affairs of large numbers of people. A set of small but beautifully carved figures in jade and

Besides their colossal stone heads, the Olmec created superb sculptures of the human form. *This lively figure, thought to be a wrestler, is in the National Museum of Anthropology in Mexico City.*

This intriguing group, carved in jade and serpentine, was unearthed in the jungle arranged in this way, and may be a clue to Olmec ceremonial life. *The figures are now in the National Museum of Anthropology.*

serpentine, found at La Venta arranged in a group, is one of the few clues to Olmec ceremonial life.

One of the distinguishing features of the Olmec culture is a special art style, based on the representation of fiercely naturalistic creatures: snarling jaguars, striking serpents, a monkey-eating eagle and strange, weeping human figures. Sometimes these motifs are combined into a babyfaced human figure with a fanged or snarling mouth. This is probably a rain god, and is one of the first recognisable deities in the Mesoamerican pantheon.

There is no doubt about the primacy of the Olmec civilisation in Mesoamerica. Whether by conquest, emulation or trade their example of a society devoted to the worship of gods and ancestors, and ruled by an elite class, spread north into central Mexico and south to Costa Rica. The Olmec even appear to have reached the Pacific coast of Mexico. In 1966 a series of vivid cave paintings was found in dry hills back from the coast, showing an Olmec ruler, wearing jaguar leggings and arm coverings, standing over a captive. Even more significant was a great red-painted snake wearing a crest of green feathers. This is the earliest known representation of the feathered serpent, the most ancient and important of Mesoamerican deities.

Although there is no direct record of it from the Olmec region, evidence from later cultures suggests that the Olmec may have developed the first written script in the Americas, as well as the complicated calendar, based on a 365-day year, that was to be elaborated by later groups. But despite their all-pervading influence the collapse of the Olmec, when it came, was apparently quite sudden. The reasons for this are a mystery, as is much about the lives of these powerful but shadowy jungle dwellers. We know more about their immediate successors, because they left a more descriptive record of their civilisation, one of the most spectacular in all the Americas—the Maya.

The Mayan culture began on the Pacific coast of southern Mexico and Guatemala, but soon spread into the southern highlands and finally across

the vast flat peninsula of Yucatan. At the height of their civilisation—the 'classic' phase which lasted for six centuries, from AD 300 to 900—the Maya developed the most sophisticated of all the American societies. They were the only fully literate people, and were great innovators in science and mathematics. Like the Arabs in the Old World, they invented the zero to assist in numeration. Their calendar recorded dates of historic events which can now be reconciled with our own calendar. Their art in all its forms is rich and highly imaginative. And the supreme expression of all those skills, their enduring and unforgettable monument, is their architecture, which survives today in majestic ruins all across Mesoamerica.

Most of the Maya were farmers, and from an early date they began to intensify their corn agriculture by building chinampas or raised fields in swampy areas (a technique which was to be brought to an amazing level of sophistication by a later culture, as we shall see). This improved productivity helped to increase the population and expand the Mayan influence. More people began to live in towns and small cities, and the farmers often had to lend their labour to help build and maintain the great stone buildings—pyramids, palaces, temples, aqueducts and ball courts. Artists adorned the public buildings with carved stone panels and painted stucco mouldings, which must have presented a dazzling spectacle, rising from the green jungle.

The craft workers in the Mayan centres created luxury items for the elite, and trade goods which circulated over a wide area of Mesoamerica. They produced wood carvings, elaborate pottery sculptures, ornaments in bone and shell, and delicate figures in obsidian and jade. Mayan books were made from folding screens of bark paper, covered with a coating of plaster. On these the scribes wrote and painted, in gorgeous colours, the complex mythologies and astronomical calculations that most concerned the Mayan leadership. Mayan art, especially in their murals, reflected the colourful associations between humanity, the gods and the many layers of the universe.

Much of our knowledge of the Maya comes from their often stylised written records of gods and other mythic beings. But occasionally we can get closer to real people from this period, more than 1000 years ago. In 1952 a Mexican archaeologist, Alberto Ruz, made one of those discoveries which occasionally communicate the excitement of this kind of research

The Maya were the only wholly literate people in the Americas, and their art records a culture of great style and character. *This pottery figure—perhaps an official making a speech—shows the Mayan love of costume.*

into the human past.

Ruz was working in the Temple of the Inscriptions at Palenque, on the edge of the southern highlands. Palenque is built against a dark backdrop of jungle on a magnificent site at the edge of the escarpment, above the grassy plain that stretches north into Yucatan. It is one of the finest and best preserved of all the Mayan centres, and the Temple of the Inscriptions is the tallest and most striking of all the buildings. The temple itself is perched on top of a steeply stepped pyramid, more than twenty metres high.

Ruz was studying the inscriptions inside the temple when he noted a double row of holes, filled with stone stoppers, in one of the slabs forming the floor of the temple. It seemed to Ruz that the only function of the holes would be to enable the slab to be lifted with ropes, so that is what he did. Underneath he found a staircase leading down into the heart of the pyramid, but filled to the top with stone rubble. It took Ruz four years to clear out the rubble, following the stairway down. Deep inside the pyramid the staircase turned 180 degrees, and continued down. Eventually, at a level just below the ground surface of the plaza in which the pyramid stands, the staircase gave way to a passageway, blocked by a huge triangular stone slab.

When Ruz moved the slab to one side he found himself looking into a large funerary crypt, nine metres long by seven metres wide, with a high vaulted ceiling which had grown stalactites. Around the walls were huge

The most enduring expression of Mayan culture was their architecture, visible now in astonishing ruins rising from the jungles of Yucatan and Guatemala. *The pyramid beneath the Temple of the Inscriptions at Palenque contains the tomb of the great ruler, Pacal, who died in AD 683.*

The men of Tenejapa still wear their traditional black poncho and tasselled hat on market day. *The language of these people goes back to Mayan times, and is incomprehensible to Spanish-speaking Mexicans.*

stucco figures of men in ancient costume—perhaps the nine lords of the night in Mayan cosmology, or possibly relatives of the deceased. In the centre an enormous honey-coloured block of sandstone, deeply incised with complex designs, formed the lid of a large sarcophagus. It took great efforts with block and tackle to lift the stone slab, but under it Ruz found an astonishing display. The skeleton of a Mayan ruler was almost buried in a treasure-trove of carved jade—masks, necklaces, beads, rings, figures, and ear spools. The dead man held a carved jade piece in each hand, and another had been placed in his mouth.

Careful study of the inscriptions on the pyramid, the temple and the tomb itself has shown that the man buried in the sarcophagus with such regal trappings was the lord Pacal, the greatest of all Palenque's rulers. He came to the throne when he was twelve years old, and died in 683 at the unusually ripe age, for those times, of eighty. Pacal had lived long enough to have the pyramid built in his lifetime as a sepulchre for himself, as the pharaohs of Egypt had done. The temple on the summit was for his subjects to worship his memory.

Not long after the reign of Pacal, Mayan society began to decline, and by about 900 it had collapsed, in what has been described as one of the most profound cultural failures in human history. City after city stopped building the great monuments. Trade came to a standstill. The vital astronomical calculations and the calendar were abandoned. There is evidence of warfare in some places. In any case the people seem to have just left the cities. The jungle swallowed up the temples, the palaces and the ball courts. It was as if the Maya had simply ceased to exist. The precise reasons for this decline are still not known, and suggestions range from an ecological disaster, such as a climatic change which affected agriculture, to a socio-political revolt by the masses against their rulers.

In fact, the culture of the Maya did not totally disappear. Many elements of Mayan tradition, including language, crafts and corn agriculture, survived and have continued right down to the present in small villages tucked away in isolated parts of the southern highlands. Despite the conquest of this region, Chiapas, by other Mesoamerican societies, and later by the Spaniards in the sixteenth century, life has continued virtually unchanged in villages like Tenejapa, a small market town in a deep valley.

We went to Tenejapa particularly to see the weavers, part of a revival of Mayan crafts in Chiapas. Here, in the Mayan tradition, the women are the weavers and cloth-makers. They spin the cotton and weave and embroider basic clothes for all the people, especially the men's rectangular poncho and the women's huipil, or shawl. Both these garments are virtually unchanged from the time of the Maya.

In their little cooperative we watched two women use the backstrap loom, one of the oldest weaving devices in the Americas. To weave in this way, one end of the warp is attached to a doorpost or any handy anchor point, and the weaver keeps the threads stretched tightly by leaning back inside a loop at the other end of the warp, while she uses both hands to shuttle the weft threads back and forth. We watched the women using the same motifs—stylised frogs, birds and geometric symbols—that appear on the walls at Palenque and other Mayan centres.

The local cemetery, too, is a link with the past. With its wooden crosses it looks like any other Christian cemetery, but the placement of the crosses, and the strange wooden planks placed at angles on the earthen mounds over the bodies, conform to a much older religious practice. The Catholicism brought by the Spaniards has taken deep root in many parts of Mexico, but here it seems more like a veneer over much more ancient beliefs.

Market day in Tenejapa also reflects the mix of past and present. The street stalls sell many traditional foods—cornbread, tortillas, peanuts, avocados, and chillis. Women's colourful huipils are on sale, too, as are the black woollen ponchos used by the men to keep themselves warm at night in the mountains. Men promenade in pairs up and down the streets in their ancient costume of short black skirt and poncho, a decorated belt, leather sandals and a conical hat festooned with brightly coloured ribbons. The local band sitting in the plaza plays sentimental, vaguely melancholic songs on a combination of modern guitars and a traditional wooden harp, while women, shawled in black, sit outside the church and listen.

Even the drinking among the men—a familiar scene on market day in many parts of the world—has ritual significance, and is ordered precisely

On market day in Tenejapa the local musicians play in the village plaza. *The home-made wooden harp is a traditional instrument in Chiapas, the southern highlands of Mexico.*

The women weavers of Tenejapa, a village in the southern highlands, keep the Mayan traditions alive with their use of designs and motifs from 1000 years ago. *These women are using the backstrap loom, which has been known in the Americas for thousands of years.*

as it was hundreds of years ago. Men from a number of surrounding villages join the men of Tenejapa, dressed in their best costumes and tasselled hats. Arranged in a circle, they drink steadily from vessels made from a large cow's horn, carved and silver mounted. After each pull they pass the flask to the man on their left. The drink is pulque, a strongly alcoholic brew made from cactus, and as the morning passes the men grow more boisterous, and their archaic local language, already unintelligible to Spanish-speaking Mexicans, becomes even more impenetrable.

The mountain people of Chiapas are living successors of an ancient culture. In them the Maya, unlike the Olmec, have a continuity of at least some aspects of their society. But Mexico saw many other civilisations rise and fall alongside the Maya without leaving much trace—except, in some cases, accounts of their unusual pastimes.

One was the Veracruz culture on the Gulf of Mexico, next to the Olmec heartland. Its people took the primitive ball game invented by the Olmec, and elaborated by the Maya, and raised it to a high level, both as a contest and a spectacle. The game was played by two teams on a rectangular court with sloping sides, surrounded by spectator galleries. Each team had to drive a heavy solid rubber ball through their opponents' stone ring, mounted on the end wall like a basketball ring. Games were played before large crowds, no doubt attracted by the overtones of death. The winning captain was entitled to great privilege, and a player who scored a 'goal' could claim the jewellery of the spectators. But the captain of the losing team was sometimes sacrificed on the spot.

In central Mexico the Toltec erected huge temples in their capital of Tula, their roofs supported by magnificent stone figures which still stare out across the arid hills. The Toltec were a warlike people who displayed the skulls of the defeated in racks next to their temples, which were decorated with carved figures of eagles eating hearts, and prowling jaguars and coyotes. A wall, still standing, carries a repetitive carving of human skeletons being swallowed by rattlesnakes. The pervading myth of the feathered serpent also recurs throughout Tula iconography.

The vast ruins of Teotihuacan, near Mexico City, mark the site of the greatest early city in Mesoamerica. *The axis of the city was the Avenue of the Dead, on the right. In the distance looms the Pyramid of the Sun.*

The most remarkable of all these lost cultures, however, is the one represented by the great deserted city of Teotihuacan, spread across the flat plain forty kilometres northeast of Mexico City. This was the largest and most impressive city in all the Americas before the arrival of the Europeans. It was a thriving metropolis 2000 years ago, at the time of Imperial Rome and Chang-an, the Han dynasty capital of China, but its twenty square kilometres of pyramids, temples, palaces, storerooms and houses was bigger than both of them put together.

The axis of the city is a broad thoroughfare, the Avenue of the Dead, which runs north and south for nearly four kilometres. The ceremonial centre of the city is dominated by gigantic stone structures that can be seen from a long way off. These are two enormous stepped pyramids, built to worship the deities of the sun and the moon. The Pyramid of the Sun is seventy-five metres high, and placed at the centre of the Avenue of the Dead. The Temple of the Moon, slightly smaller, stands on a low hill at the northern end of the great avenue. Facing the southern section of the avenue is the imposing citadel, where it is thought the priests and rulers lived. Many of the city buildings were originally covered with a thick layer of plaster and gaudily painted, chiefly red. There are also remnants of decorative murals in some of the temples and dwelling houses.

One of the most striking edifices is the Temple of Quetzalcoatl, the feathered serpent deity who will, according to legend, return one day from the east to rule Mexico once more. This is a stepped pyramid, from the front of which project the thrusting and realistically carved stone heads of feathered serpents and fire serpents. These represent the fundamental existential conflict between the greenness of life, sustained by water, and the hot, fiery deserts that surround the valley.

It is estimated that the population of Teotihuacan may have been as high as a quarter of a million—truly enormous for that period in human history. Although the land around the city was intensively cultivated it seems unlikely that such a large population could have been self-supporting, so there must have been extensive trade with other areas to bring in food. To pay for such imports, Teotihuacan became a busy craft centre. Workers made distinctive pottery, painted ritual objects, incense burners, tripod vessels, naturalistic animal sculptures, and finely flaked obsidian tools. They also made beautifully carved and decorated funeral masks, some of which were life size, and perhaps intended as lasting portraits of the dead.

But despite these material remains the great culture of Teotihuacan people, their social and political life, remains a mystery, because they had no written language. Most of what we know about their religion and cosmology comes from the accounts of societies that preceded or followed them. It is therefore impossible to be sure what caused the sudden collapse of the Teotihuacan universe in the seventh century AD, although it is possible that accumulating environmental damage to the once fertile valley may have been a factor. There is also clear evidence that part of the city was destroyed by fire, either by invasion or internal insurrection. But

The feathered serpent, with its fierce jaws and plumed head-dress, was one of the most feared deities in Mesoamerican mythology. *This carved head is on the Temple of Quetzalcoatl at Teotihuacan.*

Teotihuacan was a great craft centre, and among the products of the workshop were funerary masks of turquoise and coral. *Masks like this were traded over a wide area—even to the Pacific coast, where this one was found.*

for whatever reason, it seems that, as with so many Mesoamerican societies, the people just abandoned the city. Weeds and dust slowly covered up the great metropolis, until it became the focus of a long program of excavation and restoration which began earlier this century.

Apart from the massive ruins—perhaps the most impressive historical spectacle in North America—one other aspect of the Teotihuacan culture did survive. Many of its gods were reborn in the Valley of Mexico several hundred years later, in the last and in many ways the greatest civilisation in the Mesoamerican cultural succession.

The capital, Tenochtitlan, stood not far from Teotihuacan, and its people built many other cities in what became a great empire. This civilisation came to an end only 500 years ago—and yet not one of its great cities or palaces has survived. We can see more of the monumental heritage of the Olmec, after 2000 years of wars, looting and development, than we can of the Aztec.

The Aztec capital of Tenochtitlan today lies in rubble beneath Mexico City, the world's largest metropolis. In the Plaza of the Three Cultures you can see the ruins of the Aztec temple that once stood there, until it was torn down by the Spaniards to build a large church and monastery. This great grey stone pile still looms over the foundations of the temple, which is slowly being excavated and restored by Mexican archaeologists. The whole plaza is surrounded by tall buildings of the modern Mexican state, whose people, with an emerging sense of history, are now looking beyond their Spanish colonial heritage and finding new pride in their Meso-american roots.

The best evocation of Mexico's past is to be found in the National Museum of Anthropology. This is one of the world's great museums, where all the dazzling cultures of the past 3000 years are represented by superbly displayed statues, ceramics, decorative objects, jewellery, tools and weapons, and amazingly realistic dioramas. Here is the world of the Aztec in all its colour, bustle, life—and horror.

The centrepiece of the Aztec section is an enormous model of the ceremonial heart of Tenochtitlan, backed by a gigantic mural. Here is the sight which astonished Cortes and his men when the Spaniards marched over the mountains in 1519—a great city seemingly afloat in the Lake of the Moon, anchored to the shores by causeways. The visitors were stunned. There was nothing like it in Europe. They described it as 'another Venice', because of the network of canals dissecting the suburbs. The centre of the city was a blaze of colour, for the Aztec painted their buildings in glowing tones of red, green, yellow and turquoise. The capital was a symbol of the vitality of the Aztec nation of ten or twelve million people, at least 100 000 of whom lived here, at the centre of the empire. The vast markets were crowded with merchants, their stalls stocked with a profusion of foods and trade goods.

The city was a marvel of civil engineering. The Aztec engineers had solved the problem of salinity in the lake by damming off the greater part

The Aztec capital, Tenochtitlan, presented a dazzling sight to the Spaniards under Cortes when they entered the city in 1519. *The Aztec used bright colours in their architecture, as this model in Mexico City shows.*

The crowded markets in the Aztec capital, Tenochtitlan, sold food of every description—including snakes. *The markets have been re-created in a large and amazingly detailed diorama in the National Museum of Anthropology.*

of it. Springs near the city maintained a freshwater lagoon around Tenochtitlan itself, enabling the people to raise fish, waterfowl and salamanders—prized edible amphibians. They also hunted deer and other small game around the lake, but the basis of their food supply were the remarkable chinampas, or ingenious agricultural islands (sometimes erroneously called floating gardens). These were made by digging the rich mud from the lake bottom and piling it up into an extensive network of raised rectangular fields, separated by canals. The farmers fertilised the chinampas every year with fresh mud, and the resulting gardens were so productive that they could provide Tenochtitlan's entire population with an unfailing supply of corn, chillis, squash, beans, fruits and flowers.

Rising from the ring of hills to the east of Mexico City today, as they did in Aztec times, are three large volcanoes—one of which, Popocatepetl, towers to 5542 metres. Together with the earth tremors which frequently rumble through the Valley of Mexico, these constant reminders of the threat of supernatural and uncontrollable forces may have been partly responsible for the obsessive religious extremes among the Aztec that the Spaniards found when they arrived.

The Aztec had inherited their basic religious structure and their pantheon of fierce jaguar gods and snake goddesses from the daemonology of Teotihuacan and the Mayan and Olmec cultures. But their astronomy and cosmology had led them to the depressing conclusion that human existence was marked in cycles, and that their civilisation—the fifth and final re-creation of the universe—was doomed unless they could delay the setting of the fifth sun. The great calendar stone unearthed from beneath Mexico City, and now displayed in all its grim detail in the National Museum, records the four previous creations, with the sun god at the centre surrounded by human hearts. Thus was set in motion the ghastly response of the Aztec to the threat of annihilation: human sacrifice to appease the fire god, on a scale unknown anywhere else in the world.

The almost daily sacrifices, totalling thousands of people every year, were made outside the temples on top of the great pyramids in Tenochtitlan. Priests used obsidian knives to cut open the chests of slaves and captives. The beating hearts were torn out and flung into stone bowls, and the bodies of the victims were rolled down the steep steps into the plazas.

One of the stone bowls in which the Aztec priests flung the hearts of their sacrificial victims. *It was such practices which provoked Cortes and his men into the destruction of the Aztec capital.*

An obsidian knife with a decorated handle, probably used by Aztec priests to cut the living hearts out of captives offered for human sacrifice. *Human sacrifice was believed by the Aztec to be the only way of staving off the end of the world.*

But despite the ever-increasing parade of sacrifices the Aztec could see no escape from their fate, and by the beginning of the sixteenth century, when Cortes arrived, they were close to despair. Their ruler Motecuhzoma (Montezuma) was unnerved by predictions of disaster and strange portents, and obsessed by the legend that Quetzalcoatl, the great feathered serpent-king of the Toltec, would one day come back and reclaim his kingdom.

So when the astute Cortes, learning of this legend, claimed to be Quetzalcoatl, he was permitted into the heart of Tenochtitlan with his small band of troops, supported by disaffected tributaries of the Aztec state. Although no strangers themselves to blood and death, the Spaniards were revolted by the daily torrents of human blood that flowed down the temple steps, and needed little excuse to eventually raze the entire city.

With the destruction of Tenochtitlan the richness of Aztec life simply disappeared—its language, music, art and science. The population was

At weekends the people of Mexico City enjoy a surviving section of the canals and chinampas or 'raised gardens' which the Aztec built to grow corn and vegetables for their capital, Tenochtitlan. The elevated fields beside the canals are still producing food for Mexico City, after 500 years.

halved in twenty years. Perhaps six million people were either slaughtered or died from cruelty, starvation and disease. And so, with the overthrow of the last and perhaps most spectacular civilisation of the New World, 3000 years of Mesoamerican culture came to a bloody end. The Aztec predictions had come true.

Today the Lake of the Moon has dried up, and Mexico City sprawls across its bed of silt (whose basic instability is responsible for the shuddering earth movements which periodically demolish parts of the metropolis, with huge loss of life). But in one corner of the city there survives, after 500 years, a fascinating remnant of the ancient Aztec capital, Tenochtitlan. A section of the chinampas, or raised gardens, is still providing the eighteen million people of Mexico City with a substantial proportion of their corn, fruits, vegetables and flowers. Rejuvenated each season, as of old, by mud dug out of the canal bottoms, these elevated fields are as productive as ever, an enduring tribute to the enterprise of the Aztec city planners.

At the weekends the chinampas serve another valuable function, acting as a refreshing lung for the gasping citizens of the world's most overcrowded and air-polluted city. From dawn until dusk the canals are filled with brightly painted barges and boats, jammed with families enjoying the food and drinks supplied by floating kitchens, and listening to the cheerful songs of passing boatloads of mariachi.

The chinampas of Mexico City are a remarkable expression of cultural continuity, and one of the best surviving examples of the kinds of skills and ingenuities that were developed, quite independently from the rest of the world, by the people who first entered the Americas, tens of thousands of years before. That epic migration out of Asia and down to the tip of South America completed the human encirclement of the Pacific basin, from Tasmania all the way round to Tierra del Fuego. All that remained for those insatiably curious explorers was the vast expanse of the Pacific Ocean itself.

14.

The Last Horizon

The greatest wave of exploration in the Pacific basin, and in some ways the most astonishing, was the last—the discovery and settlement of the remote islands of the Pacific Ocean itself. This involved voyages under sail across unknown oceans, without charts or navigation instruments, that were so extended and so hazardous that they almost defy comprehension.

The Pacific Ocean is larger than the Atlantic and the Indian Oceans put together, larger than Europe and Asia combined. If you look at a globe you cannot see both eastern and western shores of the Pacific at once, because it extends more than halfway round the equator. The major groups of islands scattered across it are separated by distances that dwarf our usual yardsticks. Hawaii is much further from Tahiti than North America is from Europe, and Easter Island is further from any possible Polynesian departure point than South America is from Africa. And yet people found those and many other remote corners of Polynesia in open canoes. For boldness, ingenuity and courage no other adventure of the human species comes even close to it.

That move out across the Pacific followed the occupation of the lands which form its margins. By 40 000 years ago there were people in all four continents round the Pacific rim. Roving bands of hunter gatherers were operating along all parts of the Asian coastline. In the southwest Pacific, people had come down through the islands into New Guinea and Australia. The Pacific coast of both North and South America had seen human traffic, from Beringia down to Tierra del Fuego.

The next 30 000 years or so was a period of consolidation of those early explorations. In Asia, North America and South America human societies slowly developed, learning to live in the new environments, laying the foundations for the agricultural and metal revolutions that would one day produce new civilisations in those regions. In Australia, too, the colonists settled down to refine the techniques of hunting and gathering, finding enough space in that vast continent to support hundreds of different tribes

The Polynesians completed the peopling of the Pacific basin with their remarkable voyages of exploration and discovery across the open ocean.
Hawaiians taking part in a display at the Polynesian Cultural Centre on the island of Oahu.

and cultures.

Only on the eastern fringes of island Southeast Asia did that initial wave of human movement out of Asia maintain its momentum. There, from a base in New Guinea, little groups of people continued to push directly eastwards, into the islands of the southwest Pacific. They occupied New Britain, then New Ireland, and by about 30 000 years ago had reached the near end of the long chain of Solomon Islands.

Over the next 25 000 years these people, the Melanesians, with their voyaging outrigger canoes, settled scores of large and small islands to the east of New Guinea—probably as far out as the eastern end of the Solomons. Here, alone in their island world, they established the identity and culture of Melanesia. In their outrigger canoes they built up the island trading networks to exchange obsidian, stone tools, shells, fish and plant foods.

But then, a little less than 4000 years ago, a dramatic new influence made itself felt in the western Pacific. Although vague clues had begun to appear earlier this century, our first inkling of the extent and significance of the new culture came only in 1952, from an excavation on the south coast of New Caledonia. A group of archaeologists surveying the island found fragments of reddish pottery eroding from a bank at the back of the

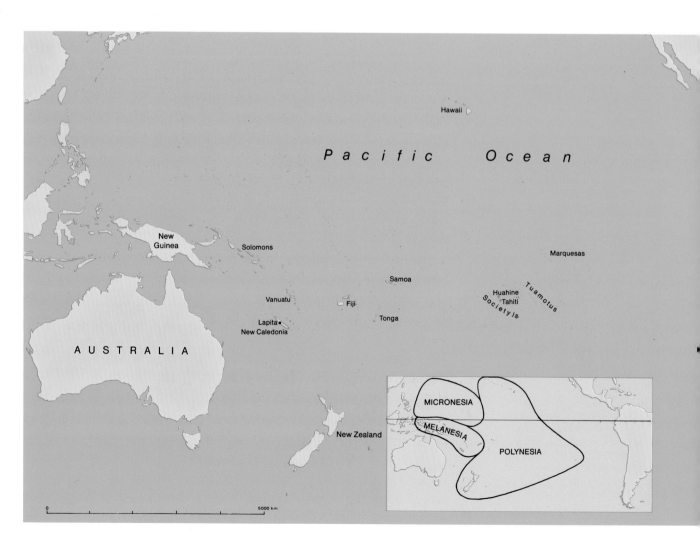

beach. Excavations of the site, at a place called Lapita, yielded scores of pieces of different pot and bowl shapes, each decorated with a different stamped or incised design. The designs were intricate, involving geometric shapes, eyes, concentric circles and shields, and had been executed with great precision. Carbon dating of material from the site suggested an occupation nearly 3000 years ago.

Since that discovery, almost forty years ago, similar pottery sherds have been found in many parts of Melanesia, and even as far east as Fiji, Tonga and Samoa. Some carried even more detailed patterns, including human faces. The type of pottery, and the culture which it represents, has been given the name of the New Caledonian site, Lapita.

What makes Lapita pottery so interesting is that although there is pottery with a similar style of decoration from the Philippines, Taiwan and Vietnam, nothing like it is known from Melanesia either before or since the Lapita pottery-making period. This lasted for about 1500 years before ending about 2000 years ago. The finds from the various Lapita sites show a steady deterioration of form and decoration in the pottery until about the beginning of the Christian era, when it disappeared altogether. The style was never revived.

So far no skeletons or other human remains have been found in any early Lapita site, so it is difficult to be certain who these people were, or exactly where they came from. In fact the question of their origins—and their descendants—is one of the most contentious subjects in Pacific anthropology. We believe, however, that there is overwhelming evidence to show that the descendants of the Lapita people can be firmly identified as the islanders we now call the Polynesians. That being so, we can make a reasonable hypothesis about where the Lapita people came from, when they entered the Pacific, and how they laid the foundations for the last great colonisation of that ocean.

The available evidence strongly suggests that around 4000 years ago pottery-making people from eastern Indonesia or the Philippines—or perhaps from the coast of Asia itself—began to move eastwards. Just why they migrated is not yet clear, but given the population growth and economic changes in that region, following the introduction of rice agriculture and the spread of metallurgy, it would not be surprising if some groups had been displaced. This would be especially true of people who, according to the evidence in the Lapita sites, did not grow rice or use metals (although in other ways they were quite sophisticated, and had stone adzes, shell tools and some domesticated plants and animals).

In a very rapid expansion the Lapita people island-hopped past New Guinea and into the southwest Pacific. Besides their undoubted skills as pottery-makers, the Lapita people were skilled boat-builders and deep-water sailors. This capacity enabled them to move out through the Melanesian islands quite rapidly, as their habitation sites show. Their skills seem to have impressed the Melanesians, because in some areas the local people adopted the Lapita terms for sailing, parts of boats, navigation and so on. (We know this because these expressions were in a language quite

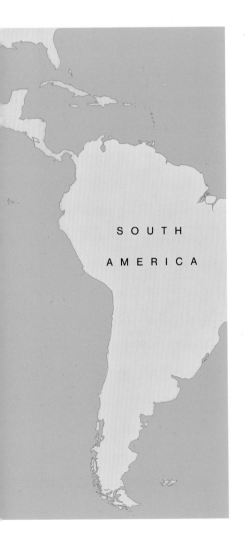

SOUTH

AMERICA

unrelated to the Melanesian languages; what the Lapita people spoke was one of a group of related languages called Austronesian, widely used in Southeast Asia.)

Sailing on past the outer, settled islands of Melanesia, the Lapita navigators began to find and establish themselves on new and uninhabited islands—New Caledonia, Vanuatu (New Hebrides) and Fiji. Then they went even further east, crossing large ocean gaps to settle the Tongan islands by 3000 years ago, and Samoa soon after. In less than 1000 years they had pushed more than 5000 kilometres out into the Pacific Ocean from their base in Southeast Asia. By this time the Lapita people had left the Melanesians far behind, and were well and truly on their own. And over the next 1000 years a new culture evolved from those Lapita pioneers—the culture of Polynesia.

In Tonga and Samoa today the origins of the Polynesians are easy to see. The people there are clearly Asian—mongoloid, with flat faces, straight

Lapita pottery, with its distinctive markings, was made by an adventurous people who came into the southwest Pacific about 3500 years ago, and whose descendants we know as the Polynesians. *This piece is from the original Lapita site in New Caledonia, and is part of the collection in the city museum in Noumea.*

The facial features and light skin of the Samoans are genetic pointers to the Asian origins of the Polynesian people. *This young boy's slightly curly black hair also reflects the admixture with the Melanesians which took place in Samoa and Fiji.*

black hair, light skin colour and the epicanthic eye-fold; all features which, as we described earlier, evolved in northern Asia as a response to the environments of the ice ages. Genetic studies, especially the analysis of the DNA molecules of the Tongans and Samoans, reflect the outward evidence, and make it clear that Polynesians are quite unlike the New Guineans or other Melanesians whom they passed on their way out into the Pacific.[1]

The Tongans and Samoans, while they were expert at fishing and collecting sea foods on the reefs around their islands, continued to depend on the plant and animal foods that their Lapita ancestors had brought with them from Southeast Asia. They had three animals—the pig, the dog and the chicken. In plants, they had coconuts, breadfruit, bananas, sugarcane and, most important of all, taro.[2] They planted taro anywhere that it would grow, which turned out to be a surprisingly wide range of environments. Taro will grow in dry ground but it can also thrive in flooded or swampy conditions where even rice could not be grown. In Tonga and Samoa taro was mostly grown in swamps or terraced, irrigated field systems.

Because of their famous sailing and navigational skills we tend to think of Polynesians primarily as seafarers. This maritime tradition was real enough, but they were also highly accomplished farmers. The systems of intensive cultivation they used with taro, for example, permitted the growth of populations that many of the islands would otherwise never have been able to support. In fact it could be said that the permanent settlement of Polynesia depended on introduced Asian animals and crops. However, the growing of plants and the raising of animals was to have a profound impact on the previously untouched islands of the Pacific, as we shall see.

One other important development accompanied the emergence of the Polynesian culture in the Tonga-Samoa-Fiji region—the appearance of the double-hulled canoe. As we explained in an earlier chapter, the Melanesians produced a genuine ocean-going canoe by removing one of the outriggers from the double-outriggers of Southeast Asia. This gave them an equally stable craft, but one that was stronger and better able to handle the waves of the open ocean. At some stage—perhaps it was introduced by the Lapita sailors—the single outrigger was replaced by a second hull.[3]

The resulting craft, with its two large hulls of almost equal size fastened beneath a single deck, and two masts with triangular sails, was stable in even mountainous seas. It was capable of carrying a sizable number of people, together with sufficient food and water for very long voyages, as well as stocks of plants and animals for colonising new islands. Thus equipped, the Polynesians were ready, 1000 years after their Lapita ancestors had arrived in Tonga and Samoa, for the next great leap forward—the voyages across the centre of the Pacific Ocean which would establish them beyond all doubt as the most adventurous maritime explorers of all time.

And so, about 2000 years ago, as the Christian era was just beginning elsewhere, the Polynesians headed eastwards again. After covering more than 3000 kilometres across open ocean they discovered the three groups of islands which were to become the heartland of their culture: the Marquesas, the Tuamotus and, perhaps 300 years later, the Society Islands, including Tahiti.

In the Society Islands the Polynesians found the high, dramatically weathered volcanic islands of Tahiti and Moorea, clothed in dense vegetation and surrounded by coral reefs and blue lagoons. The climate was gentle and the soils were rich, ideal for growing crops, especially breadfruit. And so great areas of land were cleared for planting; the initial numbers of people rapidly increased; and the Polynesian style of society, strongly hierarchical, with chiefs ruling over commoners, rapidly evolved into its most elaborate form.

The chiefs lived and worked in grand houses, usually oval in plan, with finely thatched walls and curved roofs. The floors were covered with woven palm mats. Priests conducted ceremonies on large platforms made of coral or volcanic rock, usually erected at the edge of a lagoon but occasionally high up on a mountain side. In their ceremonial life the Polynesians worshipped their ancestors, as well as important gods responsible for war, procreation and agriculture. Among the ceremonies were some that were rather grim—including human sacrifice.

There was also a great deal of warfare between islands, largely caused by overpopulation, and to carry their warriors into battle the Polynesians built extraordinary versions of their voyaging canoes. They stripped them of their masts and sails, built an elevated platform towards the bows, and

The Society Islanders evolved a stratified society ruled by chiefs and priests, who performed ceremonies on large platforms built of black lava. *This platform is on the island of Huahine, near Tahiti.*

The twin-hulled voyaging canoe, which enabled the Polynesians to make ocean crossings of many thousands of kilometres, evolved from the outrigger canoe of Southeast Asia. *The Hokule'a, anchored in a bay in Tahiti, was built to the ancient pattern, and successfully retraced the original voyage of discovery between Tahiti and Hawaii.*

The major foods taken by the Polynesians on their voyages of exploration were introduced into the Pacific islands from Asia— bananas, coconuts and breadfruit (*front*). *The grapefruit is a later development of Asian citrus, produced in the West Indies.*

added elaborately carved and painted stern pieces. Fleets of such craft were paddled into the attack, directed by priests standing on the platforms, until they crashed together and the warriors joined in close fighting with clubs. Smaller outriggers accompanied the warships to bring back the dead and wounded. (Captain Cook saw such a fleet assembling at Tahiti in 1774 for an attack on nearby Moorea; it included 160 warships and 170 smaller craft, and Cook estimated the total personnel involved to be 7760 men.)

But the Polynesians never gave up their great voyaging canoes, and about 1500 years ago they set out from the Society Islands and the Marquesas on their most astonishing voyages of all.

The precise reasons which propelled the Polynesians into the open Pacific towards places which they could not possibly have known existed— their next landfall was more than 4500 kilometres away—we will never know. Perhaps it was overcrowding, or warfare, or even a series of natural disasters, such as volcanic eruptions or tsunami, tidal waves. One interesting suggestion, by two American scientists, is that the discovery of islands inevitably led to the expectation of more islands—what they called 'auto-catalysis', or successful explorations catalysing further explorations.[4] Whatever the reasons, they were not whims or casual adventures, or chance. A long-range vessel stocked with food plants and domesticated animals, with a company of men, women and perhaps children, suggests a planned voyage of colonisation and the expectation of a new life in a new land. Here, on the broad Pacific, there was nothing to be found by an unprepared canoe swept off course while en route to a known and inhabited island.

They sailed by the sun and the stars, reading the waves, the winds and the flights of birds to maintain their chosen direction. We must assume that they headed north and east and south, seeking a dot on the horizon that would be their new home. Most of the voyages must have ended in tragedy, as food or water ran out, or storms sank the canoes. But some, we know, sailed on and finally made landfalls. The first of them was made, probably just before AD 500, by one canoe, or perhaps a small group, that had set off north from the Marquesas or Tahiti.

That first glimpse of the islands of Hawaii must have been an emotional experience for the voyagers, after weeks or months at sea. Their relief would soon have been replaced by amazement at the monstrous active volcanoes, disgorging sheets of molten lava that flowed into the sea, hissing and steaming, leaving a landscape covered with a blackened crust. It is no wonder that the new arrivals soon set up Pele, the goddess of volcanoes, to worship and appease.

The new arrivals in Hawaii found the chain of islands heavily forested, some with remarkably wet and dense rainforest. The Polynesians began to clear the forest to plant their crops—breadfruit, sugarcane, bananas and dry-soil taro. Island after island was brought under cultivation as the population increased (perhaps by the addition of later arrivals, since it is quite possible that some canoes made the return journey to Tahiti with news of the discovery of Hawaii).

The Hawaiians terraced valley slopes, and in places they cut wide stone-lined channels along the slopes to irrigate the growing crops. On several islands they cleared the upper slopes of extinct volcanoes for dry-land farming, which relied entirely on rainfall. Loose rocks were picked up and used to make retaining walls and boundary fences, which also helped to reduce moisture loss caused by high winds. On the west side of the big island, Hawaii, there are traces of these field systems covering more than 200 square kilometres. Today the slopes are used for grazing horses and beef cattle, but from the air the pattern of rock walls is still clearly visible, indicating the vast scale of the Hawaiian food-growing systems. Round the coastlines the people developed another method of food production, by constructing stone-walled fish ponds. Some covered hundreds of hectares, and had wooden grill gates that let small fish in, but once they had grown to edible size they could not get out.

Over the 1300 years of their undisturbed existence the Hawaiians managed to feed a surprisingly large population. (When Captain Cook called the first time, in 1778, there were at least a quarter of a million people in the islands.) The cost in environmental degradation was, however, very high. The land clearance caused severe erosion of the soft, rich forest soil, leaving scoured rocky slopes which are still visible over large areas of the islands. The dogs and pigs that escaped from the villages, and the rats which had accompanied the first arrivals in their canoes, combined to destroy the habitats of many native birds, and dozens of species became extinct.

There were changes to Polynesian society, too. The ceremonial rock platforms became bigger and more elaborate. Some were enclosed within walls, and had living quarters nearby for the priests. One of the most spectacular sites is at Honaunau on Hawaii, where a high rock wall extends across the base of a promontory. Here, in the so-called City of Refuge, those defeated in warfare or threatened with severe punishment

The Hawaiians evolved elaborate rituals and costumes and a rigid division between chiefs, who owned all the land, and the people, who worked it. *A drawing of a Hawaiian ruler, made during one of Captain Cook's visits to the islands.*

Like all Polynesians, the Hawaiians were expert wood carvers, and made large images of their gods. *These carved figures are in the restored Temple of Refuge on Hawaii, the biggest of the islands.*

for crimes could retreat for a period and live with the priests, later to return to the outside world without fear of molestation. Today the area has been restored, and is a national park.

The greatest changes, perhaps, were in the social structure. The original settlers had probably been closely related, owning their land together, with a chief who was one of them. But by the time the Europeans arrived Hawaiian society had become one of the most stratified in the Pacific. There were now two distinct classes—the chiefs who owned all the land, and the commoners who worked it. When Captain Cook returned from Bering Strait in 1779 he met a number of chiefs, and was particularly impressed by a young man named Kamehameha. Soon after Cook's death Kamehameha began the final restructuring of Hawaiian society, by setting out to conquer and control all the islands himself. By 1810 he had achieved his aim, and had made himself King of Hawaii.

The Polynesian culture of Hawaii was swept away over the next 100 years, but Kamehameha has not been completely forgotten. His great house facing the ocean on the big island is preserved—in the grounds of a luxury hotel named after him.

As their population grew the Hawaiians cut down more and more forest to grow food, eventually denuding large areas of the islands. *These ridges show the extent of the Hawaiian field systems, abandoned more than 100 years ago.*

Among the groups of Polynesians who set off in all directions in their great canoes from Tahiti and the Marquesas were some who must have headed south and southeast. This is difficult now to understand because, although those voyagers could not have known that they were heading into the emptiest quadrant of the Pacific, they must have realised that they were leaving the warm waters of

Towards the end of the statue-carving period on Easter Island the figures were given topknots made from red stone. *All the figures were knocked over during wars between the islanders; this row was re-erected by archaeologists.*

the tropics for colder and less forgiving seas. And yet we know that some groups took that extreme risk, because at just about the time that their fellow voyagers reached Hawaii they found a tiny, rocky dot in the southern ocean, halfway to South America, that we know now as Easter Island.

This small island is only twenty kilometres long, and not very high. It lies 2000 kilometres from the nearest habitable land, Pitcairn Island, and the Polynesian voyagers who found it must be considered remarkably lucky. However, they not only survived weeks or months at sea and made safe landfall, but began there one of the most fascinating of all the Polynesian experiments in island society.

The men and women who arrived on Easter Island brought with them the well-proved Polynesian settlement package—animals and plants, tools and implements such as shell and bone fish hooks, and, most importantly, techniques for survival by adaptation. They had yams and taro, bananas and sugarcane, even the mulberry tree for making tapa cloth. They had breadfruit and coconuts, too, but the climate proved too cool for these and they eventually disappeared. They may have taken their three food animals with them—pigs, dogs and chickens—but if so the first two either did not survive the voyage or else died out on the island, because the only domesticated animal they were able to keep was the chicken.

The island the Polynesians found was, like many Pacific islands, largely volcanic, although there had been no activity for a very long time, and the craters were worn down by long erosion. But the soil was good, and the entire island was covered with forest. So the new arrivals began to clear the forest in order to plant their crops. They built stone platforms, in the traditional way, and settled down to the kind of existence that the Polynesian people had established in so many of the Pacific islands.

But as time passed, and the population grew, the people of Easter Island embarked on an extraordinary course which was to set them apart from all other Polynesian societies. On their ceremonial platforms they began to erect moai, or stone statues, representing the founding chiefs of different family lines. This was the beginning of a cult devoted to the deification of ancestors, which was to have disastrous consequences for the islanders.

The Easter Islanders had brought with them the Polynesian skills in stone carving, and they found a virtually inexhaustible source of material on the slopes of the largest volcano, Rano Raraku. The mountain side was made up of solidified volcanic ash, or tuff, which was relatively easy to pound away with hammer stones. Each statue was carved out lying on its back, until it was attached only by a thin strip of rock. Finally this was chipped away, and the statue was slid down the slope on a ramp of earth, restrained by ropes. At the bottom it was loaded on to a wooden sledge and hauled away to the waiting ceremonial platform, where it was finished and erected.

Two things about these statues made them unusual. One was their powerful, stylised sculptural form. The faces in particular exhibited a brooding, timeless stare from large eyes set in tall, domed heads with strange, pendulous ears. The second feature about them was their size.

The early statues were quite small and naturalistic, with lifelike faces and softly rounded bodies, just like those that have been found in Tahiti. But as time passed a competition to make bigger and more impressive statues seems to have overtaken the Easter Islanders. The figures became taller, the faces more stylised, the eyes larger, the ears longer. Then they began to have cylinders of dark red stone placed on their heads, for greater emphasis.

More and more statues were carved, and more and more trees cut down to make sledges to move them. Some were transported as far as ten kilometres to be erected. The largest to be cut out and moved to its platform stands 9.8 metres high and weighs more than eighty tonnes. There is one unfinished statue on the slopes of Rano Raraku that is nearly twenty-two metres long, and is estimated to weigh more than 250 tonnes. We can only guess how the islanders, with technology limited to wooden levers, sledges and bark ropes, could possibly have moved a colossus like this. But clearly, before it could be finished a terrible fate overtook the island and its people.

Over something like 1000 years the Easter Islanders had become totally obsessed with statue-making. Between their clearance of more and more land to feed their growing population, and to provide timber for houses and sledges and poles for moving the statues, they had totally denuded the island. There were no trees left to make canoes, so the only craft were small planked vessels and reed boats. Winds had dried out the soil, and erosion had blown it away. The population, of perhaps 15 000 people, had become much too large. Famine gripped the island and decimated the population. It was a time so terrible that starving, emaciated figures with razor-sharp ribs and staring, hollow eyes are still a major subject in the carvings produced today by the islanders for sale to tourists.

Somewhere around 1700 Easter Island erupted into warfare. Rival groups armed themselves with spears tipped with shaped flakes of the

The great stone heads of Easter Island are among the most enigmatic images in the Pacific basin. *More than 1000 statues in various stages of construction have been found on the island.*

The largest Easter Island statue was never finished, but still lies on its back, partly cut from the slopes of the Rano Raraku volcano. *It is nearly twenty-two metres tall, and would have weighed about 250 tonnes.*

obsidian glass that lies everywhere on the island. Statue-building was abandoned. (Around the Rano Raraku quarry there are 394 statues left in various stages of manufacture.) Then opposing groups began overturning each other's statues—simply knocking them off their platforms. Not one statue on Easter Island was left standing. Today the few that are upright have been re-erected by archaeologists. The remainder of the more than 1000 statues that have been found on the island are lying face down in the grass or on the rocks near the beach.

At the end of the war there was a brief flourishing of a bird-man cult in the small ceremonial village of Orongo, built on a clifftop 200 metres above the sea. Priests lived in strange, low stone houses covered with turf, and carved striking figures of bird-men on the rocks. At certain times chiefs gathered on the clifftop, and their followers competed to see who could be first to climb down the cliff, swim to an offshore bird rookery, and return with an unbroken egg.

But the island society finally disintegrated and the population crashed. When the Dutch navigator Jacob Roggeveen sighted the island on Easter Sunday in 1722 (and named it accordingly) there were only about 3000 people left. Many of those were simply taken away as forced labour when European ships began to call. In 1862 Spanish slave traders abducted nearly 1500 people to work in the mines in Peru. International protests forced the Peruvian government to repatriate them, but only a handful of survivors reached the island. Unfortunately, they brought smallpox with them. By 1877 the population was down to 110 people. The few hundred Easter Islanders left today are all descended from only fifteen couples among that little remnant of the Polynesian population.

In the wars fought between themselves by the Easter Islanders their chief weapons were spears, tipped with flakes of obsidian. *Obsidian is natural glass, ejected from volcanoes, and is plentiful on the island.*

The famine and starvation which led to the collapse of the Easter Island culture have never been forgotten, and even today are reflected in the exposed ribs of the wooden statues carved by the islanders. *These examples are in the little museum on the island.*

With its isolation, its desolate brown landscape and its eerie, staring stone figures, it is no wonder that Easter Island still generates so many myths and speculations. It has also featured in more serious archaeological studies, notably Thor Heyerdahl's remarkably courageous attempts to prove, by his drift voyage from Peru to the Tuamotus on his balsa raft *Kon-Tiki*, that Polynesia was first settled from South America.

However, as we have seen, there is now not the slightest doubt that Easter Island, like all of Polynesia, was settled from the west, by people whose origins were in Asia. The evidence of language, of genetics, of the plant and animal foods, the temple buildings, even the statues—all are directly traceable to traditions in the western Pacific, not South America.

But there is an intriguing anomaly here. When Europeans first entered the Pacific many Polynesian societies—and some in Melanesia—were growing the sweet potato, a plant that is native to South America and was domesticated there. And on Easter Island there is one temple platform made up of blocks of stone that are so smoothly rounded, and so perfectly fitted in the Inca style, that they can only have been made by someone familiar with Andean stone-working techniques.

So there has obviously been some past contact between South America and Polynesia, and Easter Island in particular. It is impossible to say whether Polynesian voyagers reached South America and came back, or whether some South Americans reached Easter Island, and perhaps other parts of Polynesia. But this was certainly long after those extraordinary Polynesians had made their superhuman voyages out of the west.

This wall of dressed stone blocks on Easter Island, in the unmistakable Inca style, is often used to support theories that Polynesia was originally settled from South America. *Obviously there must have been some cultural contact, but it took place after the initial colonisation of all Polynesia (including Easter Island) from the western Pacific.*

One of the great handicaps to the full understanding of Pacific pre-history—even of such self-evident episodes as the colonisation of the Polynesian islands—is the disappearance from archaeological sites of so much evidence of human activity, through the rotting away of artefacts made of wood, fibre, bark and other perishable natural materials. Our knowledge of even the famous Polynesian voyaging canoes is rather hypothetical, based on extrapolation back in time from the kinds of canoes that were being used when the Europeans entered the Pacific. But since the canoes being used 1000 years ago were made of materials which have a life of a decade or two under normal conditions, the chances of finding one in an archaeological site would seem to be rather remote. However, occasionally there is a discovery which defies the odds.

In 1972 a construction team was building a new hotel on the island of Huahine, northwest of Tahiti. While dredging sand for landfill the men unearthed several objects made of whale bone. They were shown to Dr Yosihiko Sinoto, an archaeologist from the Bernice P. Bishop Museum in Hawaii, who happened to be working on the island. Sinoto recognised them as early Polynesian tools, and arranged for the construction work to be held up while he carried out a rescue excavation.

The many subsequent finds, and Sinoto's analysis of them, show that on the site there was once a Polynesian village, beside a small stream. About 1000 years ago this was hit by a typhoon or tidal wave, which buried everything under sand in the stream bed. The ground remained water-logged, and this preserved a whole range of things, especially wooden objects, which would normally have disappeared. (These are now in store in special fluid at the Musée National de Tahiti, awaiting final preservation treatment.) Besides the beautiful stone adzes—some still with their wooden handles—mother-of-pearl fish hooks, shell combs for making tattoos, and bone and tooth pendants, there were two groups of wooden objects which have enormous significance in the story of the peopling of the Pacific.

The largest and most important objects taken from the waterlogged soil were parts of a large voyaging canoe: a mast twelve metres long, a steering oar with the adze marks still fresh on it, and several of the wide planks, up to seven metres long, that were used to build up the sides of the great hulls. Clearly, the site had been the location of a canoe-making workshop. Here, for the first time, was material evidence that the Polynesians had built large canoes that were capable of long ocean voyages, to designs and dimensions that correspond very well to those predicated by archaeologists.

The other wooden objects were much smaller, but they filled an equally large gap in another part of the Pacific story. They looked rather like flattened clubs, with a sharp edge. Until this find, they had never been reported from Tahiti, or in fact from any other part of Polynesia—with one single exception. The Maori of New Zealand, one of the most powerful Polynesian people, did use implements like these as hand-held stabbing weapons, and they called them patu.

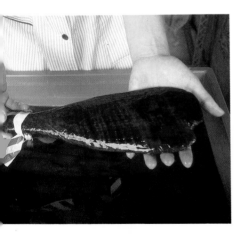

This 1000-year-old wooden patu, discovered on Huahine in the Society Islands, is the first example of the Maori type of fighting club ever found outside New Zealand. *The discovery confirms the long-held belief that the Maori came from somewhere near Tahiti.*

267

It had always seemed likely that the Maori originated somewhere in the Tahiti area of eastern Polynesia, even though their islands lie far to the southwest, much closer to Tonga and Samoa. The Maori language, their tradition of tattoos, and their style of wood carving and canoe decoration all suggested it. But there had never been any way of showing a direct link—until the Huahine excavation. The unmistakable patu and tattoo combs found there finally removed the question mark.

If you look at a map of the Pacific Ocean it seems quite extraordinary that New Zealand should have been the very last Polynesian territory to be found and settled. To begin with, the islands are very large, bigger in area than all the other Polynesian islands put together. They would offer a huge target for speculative voyagers. And compared to the magnitude of the distances between other Polynesian centres, New Zealand lies quite close to Tonga and Samoa. Why then did that discovery wait until about AD 900, until many infinitesimally smaller and more remote islands had been found? And why was it made by Polynesians from far away to the east, rather than by those to the near north?

There is no certain answer to those questions, but the southern latitude and oceanic environment of New Zealand may have had something to do with it. The seas around New Zealand can be extremely rough and stormy, and the winds and currents are usually very strong. It may be that the Tongans had discovered the conditions to the south of their tropical paradise, and were dissuaded from exploring in that direction. The pressures which prompted the exploratory voyages out of the Society Islands may have been more powerful, overcoming all fears of the southern ocean, and instigating voyages to the southwest as they did to the southeast, to Easter Island. And so it was that the Polynesians finally reached 'the land of the long white cloud', completing the colonisation of that enormous expanse of the Pacific Ocean, sprinkled with volcanic islands and coral atolls, that we call the Polynesian triangle.

In New Zealand the Polynesians found an environment quite different from any they had been used to. Because of the cool, wet temperate climate, the vegetation contained few plants that were recognisable. The forests, full of huge but strange trees, came right down to the coast in many parts. On the higher mountains, in winter, there was something totally unfamiliar—snow and ice.

Nevertheless the new arrivals, who had brought with them their Polynesian staples—coconuts, taro, yams, bananas and breadfruit—began to clear patches of forest to plant their crops. Almost immediately, however, they ran into problems, and had to make some drastic economic adaptations. Most of their tropical food plants could not tolerate the climate. Fortunately for them they had also brought sweet potato, and this plant, which was climatically more adaptable, became their main source of food, together with taro. Being resourceful, they also found a new food source: the root-like rhizomes of the native ferns, especially bracken. They then discovered that bracken would rapidly occupy any cleared ground—and the scene was set for environmental disaster.

The wood-carving tradition of the Maori is being revitalised in many craft centres in New Zealand. *Here a carver works on a large figure for display in front of a community building.*

The elaborate facial tattooing of the Maori is a strong link with the Society Islands, where similar markings were popular. *The carving also illustrates the strong Maori style of wood-carving.*

There is an enduring image of Polynesia, derived from the paintings of Paul Gauguin, of languid 'noble savages' living in a lush tropical garden. The Maori are often thought of in a similar setting of idyllic forests of tree ferns and creepers, watered by splashing waterfalls. The fact is that in New Zealand, as in Hawaii and Easter Island, the end result of the Polynesian occupation was a wholesale and irreversible degradation of much of the landscape and wildlife.

The Maori, who landed first in the North Island, cleared huge areas of forest for food growing, and set fire to much more to promote the growth of bracken fern. This stripping of the land exposed the topsoil and led to serious siltation of the rivers and pollution of coastal shellfish beds and breeding nurseries for fish. Voracious rats, which had arrived in the Maori canoes, spread rapidly, and decimated many kinds of ground birds. The new colonists had brought the dog with them as a food animal, but do not seem to have brought pigs or chickens (or if they did, they soon died out). They therefore exerted very heavy hunting pressure on the native animals. Whole colonies of breeding seals were wiped out, particularly on the North Island. Some three dozen species of birds rapidly became extinct, including a dozen different kinds of flightless moas, which proved easy prey for the hunters. Some of the moas that disappeared were gigantic birds, much bigger than the ostrich or emu.

But while the landscape of New Zealand suffered it was also highly productive, and the Maori population grew. By the time Abel Tasman arrived in the seventeenth century there were probably 100000 people living in the two islands. They had developed great artistic traditions, especially in carving wood, shell and the hard, amber-like fossil gum of the kauri pine dug out of the forest floor.

By this time, however, competition for marine resources and good agricultural land had increased dramatically and, just as in other countries where this happened, people became aggressive and warlike. The Maori tribes began fortifying and defending the land they needed for their crops. In the North Island the defensive settlements, called pa's, often resembled forts, with walls of earth and timber surrounded by ditches. Some contained pits where food was stored in winter, but they became defended retreats if the community was attacked. Extinct volcanoes made excellent sites for pa's, and some of the largest and most highly developed of them were the large cones that rise above the suburbs of the city of Auckland. Their terraced slopes and highly visible pits and defensive works are a potent reminder to modern New Zealanders of their Polynesian heritage.

And the Maori heritage, unlike that of many Polynesian societies, is still very much alive. The population is growing and languages and craft skills such as wood carving are being not merely maintained but expanded. Maori culture is an important ingredient in the New Zealand education system. The Maori, like all Polynesians, are great seamen, and they still build the great war canoes—up to thirty metres long, cut from a single tree trunk—that can carry seventy paddlers and thirty warriors. The warlike side of the Maori has been somewhat curtailed—but every rugby international sees the New Zealand team, the All Blacks, demoralise their opponents before the kick-off with the haka, a genuine Maori war cry.[5]

With the increasing willingness of New Zealand governments to recognise Maori land rights, based on the Treaty of Waitangi with the British Crown in 1840, the Polynesians seem to have a more assured future than many of the original colonists of the Pacific basin.

With the settlement of New Zealand the Polynesian triangle was complete—a colossal achievement of exploration and discovery. And it conveys a measure of the resourcefulness of the Polynesians to reflect that their climactic contribution to the occupation of the Pacific Ocean, from the coast of New Guinea to Easter Island, and from Hawaii to New Zealand, was accomplished with a wholly stone age technology. While other parts of the world had enjoyed the use of metals for thousands of years, these now essential materials were unknown to the Polynesians. They achieved all their improbable conquests with a technology limited to stone, bone, obsidian, shell, wood and other plant materials.

But that movement was itself only the final stage of a series of human movements, settlements and adaptations to hundreds of new environments around and across the Pacific basin that had been going on for more than a million years—an explosion of human energy that began in Asia and flowed out in all directions. Since people first looked out across the Pacific Ocean the inhabitants of this hemisphere had seen extraordinary environmental changes, dramatic fluctuations in sea level, the ebb and flow of glaciers and ice sheets, the creation of new islands from under the sea, and the disappearance of giant animals. They had watched civilisations and cultures rise and fall, to be replaced or swallowed up by others as new foods, new materials and new technologies came into use.

Over that immense period of time all kinds of people moved across the Pacific Ocean under many different sails—an unbroken wave of evolution and expansion that was only interrupted when over the horizon came a new people, under a very different kind of sail.

Modern New Zealanders have a constant reminder of their country's Polynesian heritage in the ruins of the Maori pa's, or forts, built on the sides of hills and extinct volcanoes. *This pa is one of the extinct volcanoes that rise within the city limits of Auckland.*

15.

The Pacific Century

If, as we suggested at the beginning, the West was lulled by its technological superiority over the past 500 years into an indifference to the Eastern Hemisphere, that attitude has certainly begun to change. The last few decades have seen the re-emergence of the Pacific peoples as a new force in world affairs. The aggregation of the human energy and industrial enterprise of the Pacific basin, led by the two giants, China and Japan, is tugging at the centre of gravity of political, economic and cultural affairs, for so long located in mid-Atlantic. The world is entering what shows all the signs of being the Pacific Century.

After a very long period of decline, colonisation and stagnation the new Pacific began to stir in the middle of the twentieth century, after World War 2, with the move to independence of a score of new nations, including Indonesia, the Philippines, Malaysia, Papua New Guinea and the Oceanic states.

Some historians may argue whether the new era began the year the French left Indochina, the day China was admitted to the United Nations — or even the hour that the British army surrendered Singapore or the second that the bomb exploded over Hiroshima. Others may pinpoint a more recent but perhaps equally significant development: the realisation by some Pacific nations that have been created in their present form by people of European stock — Australia, Chile, Canada, New Zealand — that they are not simply outposts of the West, but multicultural communities whose future lies very much in the Pacific basin.

Part of that new awareness has been influenced by the present realities of geography and economics. Part of it, however, is a growing appreciation of the deep roots of the first colonisers of those territories. There is some recognition, long overdue, that the first Australians, Chilenos, Canadians and New Zealanders are part of an occupancy of the Pacific basin that began at least 40 000 years ago.

That process, of understanding that the past of the Pacific is inextricably bound up with its present and its future, is gathering momentum, as more and more non-indigenous Australians, Chilenos, Canadians and New Zealanders begin to feel a sense of belonging, not just to their own cultural heritage, but also to a Pacific consciousness that extends back into the ice ages.

And this is as true for the descendants of the original inhabitants as it is for the newcomers. For the people of Arnhem Land, the Haida of the Pacific northwest, the Maya of Chiapas, the Peruvians of the Andes and the Polynesians of Samoa, myth is acquiring new meaning. They are beginning to learn of the ancient links that unite them, and of the great movements of people that this book, and our television series, has been about.

In the immediate future, however, we believe that there are important lessons to be learned from the past of the Pacific. One of the most urgent is for us all to evaluate the basic understanding of ecology that enabled those early colonists to adapt to new environments, and to modify their practices to the conditions they found in order to manage their surroundings without destroying them. Although some groups, like the Easter Islanders, certainly went too far and severely damaged the ecology of their environments, others maintained an acceptable balance.

The Arctic peoples, for example, hunted whales along their coastlines for tens of thousands of years, but in killing only for their own use they had virtually no effect on whale populations—certainly nothing like the devastating impact of modern whaling fleets. The hunter gatherers on all continents may have been implicated in the 'ice age overkill' which drove many large animals to extinction, but after that they devised a mythology which related them to the hunted animals on which they depended. This spiritual relationship helped them to achieve the delicate balance between catches that were sustainable and those that were not. The farmers of China, Indonesia and Peru worked their soils intensively for thousands of years without the need for artificial fertilisers or chemical pesticides.

It is sometimes argued that the past of the Pacific is not relevant to the present because of population differences—that the people of those early societies had so little impact on their environment because there were so few of them. In fact we are now coming to recognise that many areas around the Pacific basin were much more heavily populated than the early historical records have indicated.

In Mesoamerica, it is suspected, the arrival of the Spaniards reduced the population perhaps by half, through smallpox, before any serious attempts were made to estimate numbers. In North America there are estimates that the pre-Columbian population there might have been as high as ten to fifteen million people. These numbers were drastically reduced by diseases acquired from the very first parties of soldiers, trappers and traders, ahead of the general spread of Europeans from the eastern seaboard, so that the latter obtained a very misleading picture of tribal populations.

In Australia, too, a radical rethinking about the size of the pre-European population is under way. Until now, archaeologists and anthropologists have tended to assume that when the British settled Australia in the eighteenth century there were perhaps 650 tribes distributed across the continent, and that the average number of people in each tribe was probably between 450 and 500. This gave a total population, at white contact, of something like 300 000 people. Recent research on the effects of introduced diseases, however, and studies of skeletal remains of formerly inhabited areas along the Murray River, make significant revisions of that total inevitable.

A recent reappraisal of the effects of smallpox, in particular, shows that the tribal people around the first white settlement in Sydney, having no resistance to introduced diseases, were severely reduced in number by a smallpox epidemic as early as 1789.[1] As many as half of them died in April and May alone. More significantly, the disease spread in all directions, across the Great Dividing Range and down the inland rivers, into populations far from the convict settlement. There was a second smallpox epidemic in Sydney in 1829, and again it spread into the aboriginal populations.

It seems likely, therefore, that by the time explorers such as Thomas Mitchell or Charles Sturt made their journeys into the interior, and reported on the numbers of people they encountered, the aboriginal population had been reduced by as much as three-quarters. In the Murray Valley the original population densities may have been extremely high— perhaps as high as in some agricultural areas in other parts of the Pacific. There is confirmatory evidence of this suggestion in studies of skeletal remains from the area, which have found physical indicators of stress and disease typical of overcrowding.[2]

One factor which seems to have influenced the setting of a comparatively low figure for the pre-European population of Australia—an area the size of the United States or Europe—has been the notion that the desert people were typical hunter gatherers, and that the population density they had adopted was therefore typical of aboriginal society in general. Today it is recognised that the desert society was not at all representative of past population densities.

We believe, therefore, that there is now sufficient evidence to suggest that the total Australian population in 1788, when the first European fleet arrived, was well in excess of one million people.

This has implications which are both intriguing and instructive. If this population density is substantiated it will indicate a much higher productivity for the Australian landscape, whose ancient, leached soils have always been regarded as particularly unproductive and in need of liberal artificial fertilisation and pasture improvements.

One way this higher productivity was achieved, it is now clear, was by the aboriginal practice of burning—what has been described as 'firestick farming'.[3] Even today in Arnhem Land, where people still follow a largely traditional existence, virtually all the country is burned over a period of a few years. Aerial photography shows a mosaic of burned patches, where aboriginal people have set fire to the long dry grass late in the dry season. As described in the earlier chapter on hunters and gatherers, this burning produces new green shoots, attracts game, and makes the country easier to hunt across and move through. It also changes the appearance of the country.

Early European sketches and paintings of Sydney and its surroundings show an open, park-like landscape of scattered trees and short grass. It has long been assumed that these illustrations reflected the propensity of the artists to Europeanise what they saw, to translate Australian landscapes into those familiar to them. It is now clear, however, that the Sydney landscape *was* in fact open and park-like, because the local tribes burned it regularly. (Much of the present Arnhem Land country bears a remarkable resemblance to those early illustrations of New South Wales.) Francisco Pelsaert in 1629 and Captain Cook in 1770 had indicated as much, with their frequent descriptions of fires and towering columns of smoke observed while sailing along the coastline of Australia. The North American tribes also burned their territories—in California, on the northern prairies, and perhaps even in the grassy swamplands of Florida—to keep them open and more productive.

The proof of these practices has only begun to emerge since regular burning ceased, with the end of traditional control of the landscape. Satellite imaging shows that vegetation patterns are changing in several parts of the world, in ways which for a long time were not understood. In Canada the northern forests are steadily expanding south, into the prairies. In Tasmania the temperate rainforest, especially the tangled, impenetrable 'horizontal scrub', is moving back into valleys and on to button-grass plains which were open when the first European surveyed the island. In places the rainforest now comes right down to the sea, where before there had been a clear coastal strip of grassy plain. The explanation is, in both countries, the cessation of burning, which had been used for thousands of years by the inhabitants to create a particular kind of landscape which suited them.

A much more convincing proof of the effects of regular burning—or rather of its cessation—is provided, in an overwhelming and often tragic fashion, by the bushfires which periodically devastate large areas of North America and Australia. These are the direct result of the 'fire exclusion' policies of modern societies in these countries. European cultures do not use or promote fire in an economic way, as the original occupants did. To Europeans, fire is a threat to life, property and domesticated animals, and to the value of forests and the amenities of parks and reserves. So they go to great lengths to exclude fire from these areas. The inevitable consequence is the development of a dense undergrowth and the steady build-up of dead wood and other fuel—sometimes for twenty years or more.

Inevitably, fire breaks out, either by human agency or from lightning strikes during summer thunderstorms. Then, instead of the low-intensity, creeping ground fire of the aboriginal, which is usually lit in the cool of autumn and does no more than blacken the lower trunks of the big trees, there is a holocaust. The accumulated fuel, once ignited, generates sufficient heat and forced draught to create a firestorm, which races through the forest, consuming entire trees and leaping fire-breaks, destroying everything in its path. In 1988 such a fire devastated a huge area of Yellowstone National Park. Los Angeles is threatened every decade or so. Melbourne, Adelaide and Hobart are still recovering from terrible fires. Sydney and the Blue Mountains are due for one. But as yet, only in Western Australia and the Northern Territory has there been any serious experimentation with fire as a management tool in forests. In most states, the policies of national park and forestry departments still do not recognise the role of frequent burning (either naturally or by design) in the maintenance of both plant diversity and animal populations.[4]

To move on from hunters and gatherers to early agriculturists around the Pacific, there are lessons to be learned from them, too. As we mentioned in the chapter on changing the menu, food markets in many Pacific countries contain an amazing variety of plants and foods, and this reflects a much greater diversity in the species under cultivation there. Modern agricultural practice has tended, on economic grounds, towards monoculture, but in areas where foods were originally domesticated from the wild there are still a great number of breeds or varieties, distinguished by differences in colour, shape, size, texture and taste. Even in our Western adoption of these foods only a very limited range of varieties has been taken out of their original areas and put into large-scale production. This is particularly true of corn, rice, potatoes, tomatoes, beans, chilli peppers and bananas—with a noticeable loss of flavour and character, despite gains in yields and handling qualities.

Perhaps the most instructive example of the consequences of reducing genetic diversity is the so-called 'green revolution'. Well-intentioned Western agricultural scientists believed that genetically tailored high-yielding strains of some plants, particularly rice, would be of great benefit to small-scale Third World farmers—even in regions where those plants had first been domesticated. In many parts of Asia and the western Pacific, however, the green revolution has been a failure (except perhaps for pesticide and fertiliser manufacturers). The new strains of rice require heavy applications of fertilisers to produce their high yields, and constant treatment with pesticides to protect them from pests. To get the most from them, farmers have to change their traditional field systems, and buy machinery of various kinds. The cost of all this often outweighs the benefits.

In many places that we have visited farmers are abandoning the wonder strains and returning to their traditional varieties (where they can still get them, that is; one of the insidious effects of the green revolution has been the loss of seed stocks of the 'outdated' varieties). The established strains of most foods had been selected over thousands of years for their suitability to the soils and climate, resistance to pests and proven reliability.

Fortunately, there is a growing awareness in Western scientific establishments such as seed banks of the vulnerability of humanity's diminishing food base. Today, just thirty crops provide ninety-five per cent of the world's nutrition.[5] As wheat, rice and corn continue to displace indigenous crops everywhere, and highly mechanised vegetable-growing industries in countries like the United States concentrate on fewer and fewer species, the security against disease and pests that is provided by genetic variability is seriously weakened.

Interestingly, one area that is receiving close attention from organisations such as the United Nations International Board for Plant Genetic Resources is Peru. When the Spaniards arrived in Peru the twelve million people of the Inca empire were subsisting quite well by cultivating about eighty species of food crops, most of them domesticated in the Andes. Villagers owned fields at different altitudes, in order to grow as wide a range of crops as possible. The threat of famine from pests or climatic effects was reduced by this strategy, which has been described as 'the agriculture of security rather than commerce'.

Many of the native Peruvian crops were lost as the new rulers from Europe forced the farmers to produce the foods they wanted, such as wheat and barley (a disruption that was to be repeated in the badly executed agrarian reforms of 1969), but much has survived, to provide a source of wonder—and encouragement—to visiting Western scientists. A small field in the Andes no larger than twenty square metres may contain six different native grains, or up to ten different native tubers, resembling yams. Market stalls display a score of different varieties of potatoes.

Of potentially greater value to the rest of humanity are the traditional food crops that are hardly known outside the Andes. Canihua (*Chenopodium pallidicaule*) is a grain that will grow at even minus four degrees Celsius. It contains fifteen per cent protein, and is as high in the essential amino acid lysine as meat. Kiwicha (*Amaranthus caudathus*) is a grain with a protein content of fifteen to eighteen per cent which can be popped like corn, processed into flakes or milled into flour and used in cakes. Quinoa (*Chenopodium quinoa*), the sacred 'mother grain' of the Inca, has twelve per cent protein, and is high in fibre, unsaturated fats, minerals and vitamins. Tarwi (*Lupinus mutabilis*) is the legume of the Andes, and has enormous potential for other parts of the world. It contains not only as much protein (forty-eight per cent) and oil (twenty per cent) as soybeans, but also alkaloids which repel insects. Andean farmers grow it at the margins of their fields to protect them, and rotate it with potatoes to reduce nematode worms in the soil. Tarwi also fixes nitrogen in the soil and can use soil phosphorus, thus reducing the need for phosphate fertilisers.

So far, however, virtually the only Andean plant to be taken up else-where is the potato—and this, in our view, is a reflection of the dominance, over the past few hundred years, of Western models for many fundamental cultural, technological and even biological values. The nine-to-five day, the five-day week and the business suit are basic parameters in an economic system designed in the West. Many Western attitudes towards biological aspects of modern life are based on models largely derived from the results of medical or pharmacological research on white (European or Caucasian) populations—regardless of the fact that these form a very small minority of humanity.

The diversity of the indigenous peoples of the Eastern Hemisphere, and their divergence from the European model, has often been overlooked, and has in some cases led to an ignorance of their biological variability, with unfortunate results. At the simple level, there is the example—obviously not widely publicised by the US aircraft industry—of a fighter plane supplied to an Asian country which killed an unusually high number of pilots because they were not using their ejection mechanism when their plane had engine failure. It was finally realised that in aircraft designed for Americans the shorter Asian pilots simply could not reach the ejection handle in emergencies.

On a broader level, there is a need to recognise that genetic differences between Pacific peoples and Europeans do exist, and may be significant in the recognition and treatment of disease. It is fairly well known that many Asian and most Pacific island people lack the capacity to process lactose, and therefore need to avoid certain milk products. It is less well known that aboriginal Australians share this genetic trait, and that some of their other unique genetic features play a role in the incidence and severity of some diseases, such as diabetes. Alcoholism is another disease in aboriginal Australians whose possible links with genetic factors is not yet understood or even seriously investigated.

On the other hand, the Pacific peoples, and especially Melanesians and aboriginal Australians, have certain biological advantages over Europeans. They have better visual acuity, and rarely ever require spectacles. More importantly, perhaps, their dark skin results from a genetic adaptation to the intense solar radiation which gives white Australians the world's highest incidence of skin cancers.

Of all the events of the past that should concern the modern inhabitants of the Pacific basin, however, there is none that compares in immediate significance with the changes in sea level which occurred during the ice ages. Their effects on the populations of Southeast Asia, and the upheavals and mass migrations that they caused, will seem minor compared to those that are threatened by the new rise in sea level.

After many thousands of years of stability, the sea level has begun to creep up. It has already risen about fifteen centimetres since the beginning of the twentieth century, in parallel with a rise in average global temperature over the same period. However, only a fraction of that sea level rise can be attributed to the melting of glaciers or snowfields. Most of it is the result of the thermal expansion of sea water. The predicted increase in global temperature of four degrees Celsius over the next fifty years will produce another rise of about fifty centimetres, also caused by thermal expansion.

Even the small rise so far this century has begun to cause anxiety in places like Venice, the city of London, and The Netherlands. When the greenhouse effect begins to reinforce the increase in world temperatures, and the effects of melting ice are added to thermal expansion of the seas, the impact on human populations around the world, and especially in parts of the Pacific basin, will be of a quite different order of magnitude. A rise of only a few metres—considered quite possible within the next century—will virtually inundate the whole of Bangladesh, displacing tens of millions of people. Many more millions in low-lying areas of China, Japan and Southeast Asia will also be forced to move. Is there then to be another great series of forced migrations out of Asia, as there was 40 000 or more years ago?

While no one can accurately predict what will happen so far ahead, we are daily learning to make more accurate assessments of the past. While we were making our television series, and even during the writing of this book, new discoveries and revelations were being published about the prehistory of the Pacific and its inhabitants. With the growing world interest in this hemisphere those discoveries will certainly multiply. With new and better techniques for dating the past, we can perhaps look forward to solutions of early conundrums, as well as completely new dates for sites still to be found. It is interesting, then, to contemplate the future of the study of prehistory in the Pacific, and to consider what announcements may be made over the next decade.

We can expect a much more accurate estimate for the arrival of *Homo erectus* on the shores of the Pacific, and some clearer account of the slow transformation of those early humans into what we would classify with ourselves, as *Homo sapiens*. We may find hard evidence to help us understand and date the first movements out of Asia and into places like Australia and the islands of the southwest Pacific. On the other side of the Pacific there remains the great archaeological debate over the date of the occupation of the Americas. We believe that there will soon be a massive shift of scientific opinion towards a date of entry comparable to that for Australia, and a confirmation that both colonisations were the opposite arms of the same human expansion out of Asia.

We hope that more waterlogged island sites like that on Huahine will turn up, to reveal to us exactly what kind of watercraft the early Pacific mariners used to make their extraordinary voyages. Big question marks still hang over two vital human advances in the Pacific basin, domestication and metallurgy, and we expect that there will be increasingly early dates for rice and chickens, copper smelting and ironworking. It is also likely that there will be more evidence for trans-Pacific contact at earlier dates than have ever been suspected—thus helping to explain the spread of the sweet potato from South America into the Pacific, and even perhaps the finds of what look like Ming-period Chinese anchors off California.

For us, however, there have already been enough surprising discoveries about life in the Pacific basin, past and present, to more than fill a television series and a book. We remember some of the most remarkable examples of human enterprise and ingenuity that went into the occupation of this vast region. But unlike the Great Wall of China, or Machu Picchu, or the Mayan ruins, or the Borobudur, these are hardly known outside their immediate locality. They were—and in some cases still are—meant not to glorify their culture, but to fulfil some practical function, such as growing food.

The Lake Condah stone eel traps in Victoria are essentially simple devices, yet in their scale and concept they provide an entirely new picture of the first Australians. The 9000-year-old drainage systems in the New Guinea highlands testify to a capacity to grow crops before the end of the last ice age, and are one of the earliest known demonstrations of a revolution that would transform human society. The development of that agricultural revolution is indelibly etched into the landscape of the Philippines by the soaring terraces of stone-walled rice paddies at Banaue. On the outskirts of Mexico City the chinampas of the Aztec are still providing food, flowers and relaxation for millions of people. In Peru the giant staircases of the irrigated Inca field systems in the Colcha Valley were so efficiently built that they are being brought back into use. And in Sichuan Province in China the mighty Dujiangyan project is still doing the job it was built to do 2000 years ago, watering the rice bowl of the world's most populous nation.

What a story we have been privileged to tell!

The bare, treeless landscape of Easter Island—which was covered with dense forest when the Polynesians discovered it— is a warning from the past of the dangers of over-exploitation of natural resources. *Excessive cutting of trees denuded the island and caused soil erosion, which led to famine and the collapse of the population.*

Acknowledgements

The authors would like to acknowledge the assistance and cooperation of the following individuals, institutions and organisations in the making of their television series and in the preparation of this book:

ARGENTINA: Inez Segarra, Buenos Aires. AUSTRALIA: Wal Ambrose, Australian National University; Chris Ballard, ANU; Noel Barnard, ANU; Bathurst Island Community, Northern Territory; Peter Bellwood, ANU; Andrew Birbara (assistant cameraman); Tony Buckley (producer); Peter Clark, NSW Department of Lands, Buronga; Mike Dillon (second unit photography); Ian Dunlop and Film Australia; Ian Farrington, ANU; Judith Fox, Sydney; Lyle Hughes (second unit sound); Doreen Grézoux, Sydney; Jack Golson, ANU; Colin Groves, ANU; Phillip Jones, South Australian Museum, Adelaide; Rhys Jones, ANU; Michael Leigh, Australian Institute of Aboriginal Studies, Canberra; John Lovett and Lake Condah Community, Victoria; Marist Media Centre, Sydney; Dragi Markovic, ANU; Diane Moon, Maningrida, NT; National Parks Service, Tasmania; National Parks and Wildlife Service, NSW; John Oakley (co-director, senior editor); Opus Films, Sydney; Barrie Pattison, Sydney; Nicolas Peterson, ANU; Quinkan Trust, Queensland; Angela Raymond, Sydney; Grant Roberts (sound recordist); Mathew Spriggs, ANU; Tjapukai Dance Theatre, Kuranda, Queensland; Travelplan, Sydney; Pieter de Vries (director of photography); Uluru Community, NT; University of Adelaide; Weipa Community, Queensland; Helen Williams and Maningrida Community, NT; Yam Island Community, Torres Strait; Douglas Yen, ANU; Yuendumu Community, NT. CANADA: Davidson Black, Ottawa; Head Smashed In Buffalo Jump, Alberta; National Film Board of Canada, Montreal; Skidegate Band Council, British Columbia; UBC Museum of Anthropology, Vancouver. CHILE: National Historical Museum, Santiago. CHINA: Chekiang Provincial Government; Chekiang Provincial Museum, Hangzhou; China Film Co-Production Corporation, Beijing; Confucius Shrine, Shandong Province; Daye Museum, Hubei Province; Foshan Iron Foundry, Guandong Province; Guandong Provincial Government; Hubei Provincial Government; Hubei Provincial Museum, Wuhan; Institute of Vertebrate Palaeontology and Palaeoanthropology, Beijing; Jia Lanpo, IVPP; National Historical Museum, Beijing; Sichuan Provincial Government; Shensi Provincial Museum, Xian; Tsun Ko and Iron and Steel University, Beijing; Wu Rukang, IVPP; Wu Xinzhi, IVPP; Cheng Ho Museum, Nanjing; Zhengzhou Iron Museum, Henan Province. FRENCH PACIFIC TERRITORIES: Jean-Christophe Gallipaud, Department of Archaeology, Noumea; Musée National de Tahiti, Papeete. INDONESIA: Teuku Jakob, University of Gajamada, Jogyakarta; R. P. Soejono, National Archaeological Institute, Jakarta. JAPAN: Craft Guilds of Kyoto; Goza Fishing Community; Himeji Castle; Invasion Museum, Fukuoka; Clifton Karhu, Kyoto; Kyoto City Hall; Moss Temple, Kyoto; Nijo Castle, Kyoto; Ryoanji Temple, Kyoto; Yasahiro Kano, Kyoto. MALAYSIA: Lucas Chin, Sarawak Museum, Kuching. MEXICO: Gilberto Oropesa Sanchez, Mexico City; National Museum of Anthropology, Mexico City; Parc La Venta, Villahermosa; Tenejapa Community, Chiapas; Tres Zapotes Museum, Tabasco. NETHERLANDS: John de Vos, Rijksmuseum van Natuurlijke Historie, Leiden. NEW ZEALAND: Roger Green, University of Auckland; Paakaariki Harrison, University of Auckland; Te Ohaki Craft Centre, Waitomo; Waipa Kokiri Centre, Te Awamutu. PAPUA NEW GUINEA: Les Groube, University of PNG; Phillip Hughes, University of PNG; Manus Island Administration and Community; Morobe Provincial Research Committee; National Museum and Art Gallery of PNG; Swatmeri Community, Sepik River. PERU: Walter Alva and Bruning Archaeological Museum, Lambayeque; Herrare Museum, Lima; National Institute of Culture, Lima. PHILIPPINES: Jesus Peralta, National Museum, Manila. THAILAND: Pisit Charoenwongsa, Department of Fine Arts, Bangkok; Natapintu Surapol, Department of Fine Arts, Bangkok. UNITED KINGDOM: National Film Archive, London. UNITED STATES: Cahokia Mounds State Historical Site, Missouri; Thomas Dillehay, University of Kentucky, Lexington; Alan Harn and Dickson Mounds Museum, Illinois; Hotel King Kamehameha, Hawaii; Lindblad Travel, New York; Museum of the American Indian, New York; NBC News Archives, New York; Florence Naranjo, San Ildefonso Pueblo, New Mexico; National Geographic Society, Washington; Native Americans for Community Action, Flagstaff, Arizona; Ohio State Museum, Columbus; Vincent Pigott, University of Pennsylvania Museum, Philadelphia; Pu'uhonua o Honaunau National Historical Park, Hawaii; Serpent Mound National Monument, Ohio; Izumi Shimada, Harvard University, Boston; Yoshihiko Sinoto, Bernice P Bishop Museum, Honolulu; Taos Pueblo, New Mexico; Ian Tattersall and American Museum of Natural History, New York; Paddy Tatum, Kenana, Alaska; University of California at Berkeley; University of Washington, Seattle; US National Parks Service; Joyce White, University of Pennsylvania Museum, Philadelphia; Milford Wolpoff, University of Michigan, Ann Arbor; Brigham Young University, Hawaii. USSR: Nikolai Dikov, Magadan; Intourist, Moscow; Svetlana Fedoseeva, Yakutsk. WEST GERMANY: Institut Wissenschaftlichen Film, Gottingen; WDR International.

PICTURE CREDITS
Pictures are by Robert Raymond, except for the following: Alan Thorne pp. 17(u), 25, 27, 35, 39, 44(r), 45, 46, 47, 68(u), 78, 79(ul, ur), 83, 122(u), 124, 127(ur), 131, 176, 180, 181(r), 193(u), 202, 204, 214, 218(l), 227, 234(ur), 236, 240, 244, 246, 257, 260(l), 261, 269, 270; American Museum of Natural History, New York pp. 72, 73, 74, 75, 77, 91, 92, 93, 95, 96, 99, 100, 101(ul, ur), 109; John Oakley pp. 1, 6, 70, 76, 79(ll), 82, 88, 112, 113(l), 115, 116, 119, 143(u), 150, 232; Mike Dillon pp. 10, 110, 127(l), 142, 259; Pieter de Vries pp. 127(l) 132, 134; Jia Lanpo p. 23; Robert McAuley p. 28; Izumi Shimada p. 231(l); Patagonian Institute, Punta Arenas p. 223; Rijksmuseum van Natuurlijke Historie, Leiden p. 15.

280

End Notes

1: First Footsteps

1 Dubois, E. 1894 Pithecanthropus erectus, eine Menschenaehnliche Ubergangsform aus Java. *Landesdruckerei*, Batavia. Dubois returned to Holland in 1895, and was appointed professor of geology at the University of Amsterdam. He died in 1940.

2 Black, D. 1926 Tertiary man in Asia: The Chou Kou Tien discovery. *Nature* 118:733–734.
_____ 1927 Further hominid remains of Lower Quaternary age from the Chou Kou Tien deposit. *Nature* 120:954.

3 Jia Lanpo is the last surviving member of the team which began the historic excavation at Zhoukoudian (Chou Kou Tien) in 1927. In 1980 he published an excellent introduction to Chinese human fossil discoveries. In 1987 Jia Lanpo took us to Zhoukoudian and helped us to film the excavations.
Jia Lanpo 1980 *Early Man in China*. Foreign Languages Press, Beijing.

4 The reasons for this variation in the rate of change towards a more modern form are the subject of much speculation among anthropologists. One possible explanation for the apparent long-term stability of the *Homo erectus* people in Java, compared to those in China, was the fact that their tropical environment changed much less than the environment on the mainland, further north. Research in the tropics has established that plants and animals which have become well adapted to that environment often show little change over time.

FURTHER READING
Lewin, Roger 1984 *Human evolution: an illustrated introduction*. Blackwell, Oxford.
Reader, John 1981 *Missing Links: The hunt for earliest man*. Collins, London.
Wolpoff, Milford 1980 *Paleoanthropology*. Knopf, New York.

2: Casting Off

1 Chappell, J. and Shackleton, N. J. 1986 Oxygen isotopes and sea level. *Nature* 324:137–140.

2 Flood, J. 1989 *Archaeology of the Dreamtime*. Collins, Sydney.

FURTHER READING
Bellwood, Peter. 1985 *Prehistory of the Indo-Malaysian Archipelago*. Academic Press, Sydney.
Fox, J. J., Garnaut, R. G., McCawley, P. T. and Mackie, J. A. C. (Editors) 1980 *Indonesia: Australian Perspectives*. Research School of Pacific Studies, Australian National University, Canberra.
Thorne, Alan. 1980 The Arrival of Man in Australia. *The Cambridge Encyclopaedia of Archaeology*. (Ed. Andrew Sherratt) pp: 96–100. Prentice Hall/Cambridge, Scarborough.
Kirk, R. L. and Thorne, A. G. (Editors) 1976 *The Origin of the Australians*.

Australian Institute of Aboriginal Studies, Canberra.
Film Australia. 1983 *Out of Time, Out of Place*. A 45-minute film by Alan Thorne on the human fossil evidence for the origin of the first Australians. Includes film of Chinese, Indonesian and Australian sites as well as raft and computer experiments.
Opus Films, Sydney. 1976 *The Coming of Man*. A 75-minute film by Robert Raymond on recent archaeological studies of early Australians, including sites at Lake Mungo, Kow Swamp, Koonalda Cave, Devon Downs, Mootwingee, and in Tasmania.

3: Hunters and Gatherers

1 Jones, M. R. and Torgersen, T. 1988 Late Quaternary evolution of Lake Carpentaria on the Australian–New Guinea continental shelf. *Australian Journal of Earth Sciences* 35:313–324.
Blom, W. M. 1988 Late Quaternary sediments and sea-levels in Bass Basin, southeastern Australia—A preliminary report. *Search* 19:94–96.

2 Golson, Jack 1971 Australian Aboriginal Food Plants: Some Ecological and Culture-Historical Implications. pp: 196–238. In *Aboriginal Man and Environment in Australia*. (Editors D. J. Mulvaney and J. Golson) Australian National University Press, Canberra.

3 As Joseph Banks noted in his journal during August 1770, speaking of the fruits of the *Macrozamia* palm: 'We were assurd that these people eat them, and some of our gentlemen tried to do the same, but were deterrd from a second experiment by a hearty fit of vomiting and purging which was the consequence of the first. The hogs however . . . were all taken extreemly ill of indigestions: two died and the rest were savd with dificulty.' (J. C. Beaglehole [Editor] 1963 The *Endeavour* Journal of Joseph Banks 1768–1771. Vol II:115. Angus and Robertson, Sydney.)

4 Many of the now extinct marsupials, including some giant ones, were part of the fauna known to the early Australians for at least 30 000 years, and survived until the end of the ice age. Changing environments 10 000 to 15 000 years ago may have contributed to the extinction of some species, but would have been beneficial to others. It is likely, therefore, that humans, either directly through hunting or indirectly through practices such as firing the bush, were at least partly responsible for the extinction of some of the giant marsupials—the so-called megafauna. No one doubts that the Australians hunted the rhinoceros-sized *Diprotodon* or the three-metre-high *Procoptodon*, and as each animal represented a large supply of food they were clearly worth the hunting and stalking effort involved. Ironically, archaeologists have yet to find clear evidence of this hunting, or a kill site similar to those in the Americas,

where there are many such sites showing the hunting and butchering of giant animals.

5 White, Carmel 1971 Man and Environment in Northwest Arnhem Land. In *Aboriginal Man and Environment in Australia*. (Editors D. J. Mulvaney and J. Golson) pp: 141–157. Australian National University Press, Canberra.

FURTHER READING
Mulvaney, D. J. and Golson, J. (Editors) 1971 *Aboriginal Man and Environment in Australia*. Australian National University Press, Canberra.
Flood, Josephine. 1989 *Archaeology of the Dreamtime*. Collins, Sydney.
Meehan, Betty. 1982 *Shell Bed to Shell Midden*. Australian Institute of Aboriginal Studies, Canberra.
Kirk, R. L. 1983 *Aboriginal man adapting*. Oxford University Press, Melbourne.
Jones, Philip and Sutton, Peter. 1986 *Art and Land; Aboriginal Sculptures of the Lake Eyre region*. South Australian Museum, Adelaide.
Blainey, Geoffrey. 1975 *Triumph of the Nomads. A History of Ancient Australia*. Macmillan, Melbourne.
Sutton, Peter (Editor) 1988 *Dreamings: The Art of Aboriginal Australia*. Penguin Books Australia, Ringwood.

4: Into the Deep Freeze

1 Canadian scientists found seeds of the Arctic lupine in the permanently frozen burrow of a lemming, a small rodent, in the Yukon Territory. The seeds were kept and studied for some years, and estimated to be several thousand years old. When they were placed on damp paper some began to sprout. Botanists have suggested that the seeds of some Arctic plants may have survived long periods under glaciation during the ice ages, and re-established the species in their old habitats after the ice melted.
Brown, Dale 1975 *Alaska*. Time-Life International (Nederland) BV.

2 These fossil bones are evidence of the many large animals which inhabited the Northern Hemisphere before the extinctions at the end of the last ice age. Within less than 3000 years about ninety-five per cent of the large mammals of North America had disappeared. They included mammoths, bison, horses, sabre-tooth cats, bears and giant sloths.
Jacobs, Martina and Richardson III, James B. (Editors) 1983 *Arctic Life: Challenge to Survive*. Carnegie Institute, Pittsburgh.

FURTHER READING
Qiu, Pu 1983 *The Orogens—China's Nomadic Hunters*. Foreign Languages Press, Beijing.
Bruemmer, Fred 1985 *The Arctic World*. Century Publishing, London.
Birdsell, J. B. 1981 *Human Evolution*. Houghton Mifflin, Boston.

5: A Universe of Ice—and a New World

1 Chlenov, Mikhail A. and Krupnik, Igor I. 1984 Whale Alley: a site on the Chukchi Peninsula, Siberia. *Expedition*, Philadelphia. pp: 6–15.

2 From November 1986 to February 1988 *Natural History*, the journal of the American Museum of Natural History in New York, ran a series of monthly articles under the general title 'The First Americans', exploring archaeological sites and other evidence bearing on the peopling of the Americas. Some authors in this series maintain the view that there is still no positive evidence for a human presence before about 12 000 years ago, but most believe that there is good evidence for a much earlier arrival.

3 Chappell, J. and Shackleton, N. J. 1986 Oxygen isotopes and sea level. *Nature*, 324:137–140.

4 Fladmark, Knud 1986 Getting One's Berings. *Natural History*, November:8–19.

5 Guidon, Niede 1987 Cliff Notes. *Natural History*. August:6–12.

FURTHER READING
Balikci, Asen 1970 *The Netsilik Eskimo*. Natural History Press, New York.
Claiborne, Robert 1976 *The First Americans*. Time-Life International (Nederland) BV.
Dumond, Don 1977 *The Eskimos and Aleuts*. Thames and Hudson, London.
Canby, Thomas 1979 The Search for the First Americans. *National Geographic*, Washington. 156:330–363 (September).
Vesilind, Priit 1983 Hunters of the Lost Spirit. *National Geographic*, Washington. 163:151–196 (February).
Perkins, John 1981 *To the Ends of the Earth*. Pantheon Books, New York.

6: Ten Thousand Islands

1 Allen, J., Gosden, C., Jones, R. and White, P. 1988 Pleistocene dates for the human occupation of New Ireland, northern Melanesia. *Nature* 331:707–709.
Wickler, S. and Spriggs, M. 1988 Pleistocene human occupation of the Solomon Islands, Melanesia. *Antiquity* 62:703–706.

2 Groube, L., Chappell, J., Muke, J. and Price, D. 1986 A 40 000-year-old human occupation site at Huon Peninsula, Papua New Guinea. *Nature* 324:453–455.

3 Doran, Edwin 1981 *Wangka: Austronesian Canoe Origins*. Texas A and M University Press, College Station.

FURTHER READING
Bellwood, P. 1978 *Man's Conquest of the Pacific*. Collins, Auckland.
Horridge, Adrian 1985 *The Prahu: Traditional Sailing Boat of Indonesia*. Oxford University Press, Singapore.
Howells, William 1973 *The Pacific Islanders*. Weidenfeld and Nicolson, London.
Haddon, A. C. and Hornell, James 1975 *Canoes of Oceania*. Bishop Museum Press, Honolulu.

7: Changing the Menu

1 Until the 1930s it was believed that the central mountain chain of New Guinea consisted of steep and virtually unpopulated peaks. In 1933, following the discovery of large gold deposits on the lowlands near Wau and Bulolo, two Australian plantation owners, Mick Leahy and his brother Danny, set off to look for gold in river valleys leading down from the central mountains. When they climbed into the highlands they came upon the huge, fertile and densely populated Wahgi Valley, the first of several long, elevated mountain valleys that extended along the rugged spine of New Guinea. After aerial surveys they walked into the Wahgi Valley with a large government patrol and made the first contact with the people of the highlands.

2 Golson, J. 1989 The Origins and Development of New Guinea agriculture. In *Foraging and Farming: The Evolution of Plant Exploitation*. (Editors David Harris and Gordon Hillman) Unwin Hyman, London.

3 Sullivan, M. E., Hughes, P. J. and Golson, J. 1987 Prehistoric garden terraces in the eastern highlands of Papua New Guinea. *Tools and Tillage* 4:199–213.

4 Flood, Josephine 1989 *Archaeology of the Dreamtime*. Collins, Sydney.

FURTHER READING
Heiser, Charles B. 1981 *Seed to Civilisation—the Story of Food*. W. H. Freeman, San Francisco.
Simmonds, N. W. (Editor) 1984 *Evolution of Crop Plants*. Longman, London.
Harrison, S. G., Masefield, G. B. and Wallis, Michael 1975 *The Oxford Book of Food Plants*. Oxford University Press, London.
Harris, David and Hillman, Gordon (Editors) 1989 *Foraging and Farming: The Evolution of Plant Exploitation*. Unwin Hyman, London.
Tannahill, Reay 1988 *Food in History*. Penguin, London.

8: A New Cutting Edge

1 Ambrose, W. R. and Duerden, P. 1982 Pixe analysis in the distribution and chronology of obsidian use in the Admiralty Islands. In W. R. Ambrose and P. Duerden (Editors) *Archaeometry: An Australian Perspective*. pp: 88–89 ANU, Australian National University, Canberra.

2 The terms 'copper age', 'bronze age' and 'iron age' refer not to specific periods in history, but to particular technological phases in the culture of an area, beginning with the transition from a stone technology to a smelted metal technology. These phases may therefore begin at quite different dates in different parts of the world. For example, in Mesopotamia the bronze age is considered to have begun around 5000 years ago, whereas in Central Europe it did not begin until about 3500 years ago, and in South America less than 1000 years ago. The iron age in Asia Minor began around 4000 years ago, but in Britain not until 2500 years ago. In the Americas there was no iron age. This metal was not used until after it was introduced by the Spaniards less than 400 years ago.

3 A detailed and well illustrated account of the Ban Chiang discoveries, and their significance in the prehistory of Southeast Asia, was published in 1982 by the University of Pennsylvania as a special edition of *Expedition*, the magazine of the Museum of the University of Pennsylvania (24,4) Philadelphia.

4 Pigott, C. Vincent and Natapintu, Surapol 1988 *Archaeological investigations into prehistoric copper production: the Thailand Archaeometallurgy Project 1984–6*. MIT Press, Harvard.

5 The origins of lost-wax casting are unknown, but it was used in the Middle East at least 5000 years ago to make figurines or trinkets, and eventually developed independently in many parts of the world. A model of the object was made in wax and coated with clay, forming a mould. A small hole or tunnel was left in the base of the mould, extending through the clay to the wax. When the mould was heated in an oven or furnace the wax melted and ran out through the hole, and the clay became hard. Molten metal was then poured into the mould, producing an exact copy of the original wax model. To make a hollow object, such as a Dong Son drum, a thin layer of wax was applied over a solid clay core, and then covered with an outer layer of clay. The wax was melted out or 'lost', and metal poured in to fill the space that was left in the clay mould.

6 Needham's massive work is enormously detailed and very technical, and so Cambridge University Press began publishing an abridged version by Colin A. Ronan, entitled *The Shorter Science and Civilisation in China*. (Volumes I 1978, II 1981 and III 1986.) Barnard, Noel and Tamotsu, Sato 1975 *Metallurgical Remains of Ancient China*. Nichiosha, Tokyo.

7 The greatest mine of the ancient Western world—perhaps the only one comparable in size with Tonglushan—was Rio Tinto, forty kilometres inland from the Gulf of Cadiz in Spain. Mining for copper began there some 5000 years ago. Stone tools were used to extract the malachite from shallow trenches. Silver mining began about 3000 years ago, and reached its peak about 1800 years ago, when Rio Tinto provided most of the coinage for the Roman empire under Trajan. After the Visigoths overran Spain in about AD 475 Rio Tinto was lost and forgotten for 1000 years. It was reopened in the sixteenth century and is still producing copper today. Ancient workings revealed by modern open-cut operations show that although Rio Tinto was a highly developed mine under the Romans, with large underground water-lifting wheels to drain the shafts, there was nothing in the earlier workings to compare with the level of industrial technology found at Tonglushan.

8 In 1987 one of the Chinese scientists who studied the bells, Sinyan Shen, published a fascinating article about the work, 'Acoustics of ancient Chinese bells'. *Scientific American* 256:94–102.

FURTHER READING
Ambrose, Wal 1988 A bronze find in Papua New Guinea, in *Australian Natural History* 22:415.
Raymond, Robert 1984 *Out of the Fiery Furnace*. Macmillan, Melbourne.
The Great Bronze Age of China. 1980 Thames and Hudson, London.
The Metalsmiths. 1974. Time-Life International (Nederland) BV.
Tylecote, R. F. 1976 *A History of Metallurgy*. The Metals Society, London.

9: The Powerhouse

1 An interesting table of Chinese inventions, and the approximate lag in centuries before they were taken up in the West, is included in Temple, Robert 1986 *China—Land of Discovery and Invention*. Patrick Stephens, London.

2 It is not certain when the principle of the paddlewheel was taken up in the West but, in 1841, during the Opium Wars in China, the British navy used warships propelled by paddlewheels as well as sails. The Chinese navy also had paddlewheelers, and the British were reported to be amazed at the speed with which the Chinese had copied this technology.

3 The Chinese also discovered how to remove some of the carbon from this so-called white cast iron to make it less brittle and more useful—in fact, to produce steel (which contains between 0.3 and two per cent carbon, compared to the 4.5 per cent carbon in white iron). One method was to roast the cast iron tool or other implements in a furnace, in the presence of air. The oxidising atmosphere removed carbon from the surface of the cast iron, in effect giving it a steel jacket. The other method involved melting cast iron and then 'puddling' or stirring it, so that the atmosphere drew off carbon. In this way, 2000 years ago, the Chinese could make steel in large quantities and cast it like white iron or work it like wrought iron. This technological breakthrough was re-invented by Henry Cort in England in 1784!

FURTHER READING
Gernet, Jacques 1985 *A History of Chinese Civilisation*. Cambridge University Press, Cambridge.
Samagalski, Alan and Buckley, Michael 1984 *China—a travel survival kit*. Lonely Planet, Melbourne.

10: Pure and Simple

1 Even today, after centuries of attempts to devise a Japanese version of Chinese writing, the Japanese written language is still enormously complicated. It has an immense and intricate system of signs for a few dozen sounds, which in European languages are easily represented by a twenty-six-letter alphabet.

2 For a time, under the Shogun Hideyoshi, the restrained aestheticism of the tea ceremony gave way to ostentation, almost parody. In 1587 Hideyoshi announced all through Japan that he would hold a monster public tea ceremony at his great castle in Osaka, to which everyone was invited, from his richest vassals down to humble peasants. All anyone had to bring was a kettle, a cup and a mat to sit on. The affair lasted for ten days, with plays, music, dancing and displays of art treasures. On a similar occasion Hideyoshi presented guests with trays loaded with gold and silver pieces.

FURTHER READING
Austin, Robert and Ueda, Koichiro 1983 *Bamboo*. Weatherhill, New York and Tokyo.
Sansom, G. B. 1985 *Japan—A Short Cultural History*. Charles E. Tuttle Company, Tokyo.
Lowe, John 1983 *Japanese Crafts*. John Murray, London.
Living National Treasures of Japan. 1984 Museum of Fine Arts, Boston.
Maraini, Fosco 1985 *Japan: Patterns of Continuity*. Kodansha, Tokyo.

11: Flaming Arrows

1 The disappearance of tribal cultures before the rush to 'win the west' by the new settlers, and the virulent effects of the diseases that they brought with them, was so widespread and so sudden that misunderstandings about the vanishing people were inevitable. Columbus had called them Indians, but they were not from India or the Indies. They became known as redskins, but they were not red. And so misunderstandings persisted and grew, until Hollywood turned misunderstanding into myth, and created the image of the redskin firing his flaming arrows into the wagon trains that was to become deeply ingrained in the popular imagination round the world.

2 Very similar canoes were also used on the other side of the Pacific by the Tasmanians, except that they were made of rolled up sheets of fibrous bark.

3 The museum at Head Smashed In gives a graphic account of the life and times of the buffalo herds and their Piegan hunters, through re-creations of excavations, displays of artefacts and animal remains, and excellent films and photographs. The name of the site, incidentally, comes not from the condition of the buffaloes after the jump, but from an incident which, in the Piegan legend, took place in the middle of the nineteenth century, just before the great buffalo jumps ended. A young warrior wanted to watch the climactic plunge of the buffaloes from close up, and so he stood under the cliff, like a man under a waterfall, and watched the torrent of great beasts cascade past him. But the drive had been very successful, and the bodies mounted up at the foot of the cliff in a huge pile, trapping the young man against the cliff. When his people came to butcher the animals they found him with his skull crushed, and they named the place accordingly.

4 There are informative articles on the Anasazi and the history of pueblo pottery in *National Geographic*, Washington 1982, 162 (November).

5 John, Brian 1977 *The Ice Age, Past and Present*. Collins, London.

FURTHER READING
Brown, Dee 1971 *Bury My Heart at Wounded Knee*. Holt, Rinehart and Winston, New York.
Coe, Michael, Snow, Dean and Benson, Elizabeth 1986 *Atlas of Ancient America*. Facts on File, New York.
Dozier, Edward 1970 *The Pueblo Indians of North America*. Holt, Rinehart and Winston, New York.
Durham, Bill 1960 *Canoes and Kayaks of Western America*. Copper Canoe Press, Seattle.
Graybill, Florence Curtis and Boesen, Victor 1981 *Edward Sherriff Curtis: Visions of a Vanishing Race*. American Legacy Press, New York.
Harn, Alan D. 1980 *The Prehistory of Dickson Mounds: The Dickson Excavation*. Illinois State Museum, Springfield.
Irwin, R. Stephen 1984 *Hunters of the Sea*. Hancock House, Surrey. (This is one of a most informative series on North American hunters, others being *Hunters of the Ice*, *Buffalo*, *Northern Forest* and *Eastern Forest*.)
Kendrick Frazier 1986 *People of Chaco: A Canyon and Its Culture*. W W Norton and Company, New York and London.
Kopper, Philip 1986 *The Smithsonian Book of North American Indians*. Smithsonian Books, Washington DC.
MacDonald, George 1983 *Ninstints: Haida World Heritage Site*. University of British Columbia Press, Vancouver.

12: Roads without Wheels

1 Guidon, N. and Delibrias, G. 1986 Carbon-14 dates point to man in the Americas 32,000 years ago. *Nature* 321:769–771.
Delibrias, G., Guidon, N. and Parenti, F. 1988 Stratigraphy and Chronology of the Toca do Boqueirao do Sitio da Pedra Furada. *Archaeometry: Australian Studies 1988* (Editor J. R. Prescott). University of Adelaide.

2 Dillehay, T. D. 1984 A Late Ice-Age Settlement in Southern Chile. *Scientific American* 251:106–117.

FURTHER READING
Alva, Walter 1988 Discovering the New World's Richest Unlooted Tomb. *National Geographic*, Washington. 174:510–555 (October).
Goodall, Rae 1979 *Tierra del Fuego*. Ediciones Shanamaiim, Ushuaia.
Doe, Michael, Snow, Dean and Benson, Elizabeth 1986 *Atlas of Ancient America*. Facts on File, New York.

13: The Feathered Serpent

1 The avocado (*Persea americana*) is one of those plants that disperses its seeds by packaging them in a tasty, aromatic, pulpy fruit, likely to be eaten by browsing animals large enough to swallow them. The seeds pass through the animal as it wanders, and are usually deposited—complete with fertiliser—some considerable distance from the parent tree. It seems likely that the vehicles for dispersion of the avocado and many other American plants were giant ice age animals—elephant-like gomphotheres and mammoths, ground sloths, horses and other large herbivores. When the so-called 'overkill' of these animals took place after the end of the ice age many plants lost their dispersing agents, and some species became extinct, while others were restricted in range. The avocado was one that survived this problem, even though there were no longer any browsing animals in Mesoamerica large enough to swallow the seeds. The place of the former dispersing agents was taken by hunter gatherers, who collected avocados on their foraging trips into the forest and carried them back to their camps. Later these same people began to plant the avocado seeds, and thus domesticated this useful tree.
Janzen, Daniel H. and Martin, Paul S. 1982 Neotropical Anachronisms: The Fruits the Gomphotheres Ate. *Science* 215:19–27.

FURTHER READING
Coe, Michael 1987 *The Maya*. Thames and Hudson, New York.
——, Snow, Dean and Benson, Elizabeth 1986 *Atlas of Ancient America*. Facts on File, New York.
Davies, Nigel 1983 *The Ancient Kingdoms of Mexico*. Penguin, Harmondsworth.
Fagan, Brian 1984 *The Aztecs*. W. H. Freeman, New York.
Soustelle, Jacques 1984 *The Olmecs: the Oldest Civilisation in Mexico*. Doubleday, New York.
Valades, Adrian Garcia 1987 *Teotihuacan*. Editores GV, Mexico.
Simmonds, N. W. (Editor) 1984 *Evolution of Crop Plants*. Longman, London.
Coe, Michael, Snow, Dean and Benson, Elizabeth 1986 *Atlas of Ancient America*. Facts on File, New York.

14: The Last Horizon

1 Hill, A. V. S. and Serjeantson, S. W. (Editors) 1989 *The Colonisation of the Pacific—a Genetic Trail*. Oxford University Press, Oxford.

2 Other Southeast Asian plants that were important to the Polynesians were the paper mulberry (*Broussonetia*), whose bark was beaten out to make the beautifully decorated tapa cloth, and a small shrub of the *Piper* (pepper) genus, whose roots, crushed and mixed with water, produced the intoxicating drink kava.

3 In *Polynesian Canoes and Navigation*, published by the Institute for Polynesian Studies at the Brigham Young University in Hawaii, it is suggested that the Lapita people who settled Tonga and Samoa arrived in double-hulled canoes. These, it says, were probably very similar to the Fijian drua, 'a double-hulled vessel of outstanding mobility that could carry up to 300 men in addition to cargo'. The booklet continues: 'Both Tongans and Samoans adopted the drua with slight modification: it had a full covered deck, a large raised platform (with crew housing), and booms connecting the hulls amidships. With masts stepped and sails set under steerage with bow and stern paddles up to forty-five feet long, the craft were extremely stable under way at speeds in excess of ten knots'.
Thompson, Judi and Taylor, Alan 1980 *Polynesian Canoes and Navigation*. Brigham Young University Hawaii Campus, Laie.

4 Diamond, Jared. 1985 The Riddle of the Ancient Mariners. *The Sciences* (May/June).

5 The name All Blacks comes not from the ethnic character of the team but from the colour of the uniforms and boots worn by the players. In fact, although the team invariably includes some Maori members, the All Blacks are predominantly white.

FURTHER READING
Bellwood, Peter 1978 *Man's Conquest of the Pacific*. Collins, Auckland.
——1987 *The Polynesians*. Thames and Hudson, London.
Finney, Ben 1979 *Hokule'a. The Way to Tahiti*. Dodd, Mead, New York.
Haddon, A. C. and Hornell, James 1975 *Canoes of Oceania*. Bishop Museum Press, Honolulu.
Howells, William 1973 *The Pacific Islanders*. Weidenfeld and Nicolson, London.
Kirch, Patrick V. 1985 *Feathered Gods and Fish Hooks: An Introduction to Hawaiian Archaeology and Prehistory*. University of Hawaii Press, Honolulu.

15: The Pacific Century

1 Butlin, Noel 1983 *Our Original Aggression. Aboriginal populations of Southeastern Australia 1788–1850*. Allen and Unwin, Sydney.

2 Webb, S. G. 1987 A palaeodemographic model of late Holocene Central Murray Aboriginal Society, Australia. *Human Evolution* 2:385–406.

3 Jones, R. 1969 Fire-stick Farming. *Australian Natural History* 16:224–228.
Flood, Josephine 1989 *Archaeology of the Dreamtime*. Collins, Sydney.

4 Many species of Australian native plants, such as banksias, only drop their seeds after the seed containers have been scorched by a low ground fire. In the habitat favoured by banksias the ground beneath the trees is invariably deep in dry grass and leaf litter. Seeds falling on this would be trapped high above the soil, with little chance of germinating, but with a good chance of being found by seed-eating animals. Thus, although seeds are produced every year, in the absence of intense heat they are held in their containers, sometimes for years. A bushfire changes this situation completely. The banksias survive the usual low-intensity bushfire because of their thick, insulating bark, even though the leaves may be burned off and the seed containers charred and blackened. A few days after the fire the containers open and the seeds fall into a deep bed of nutrient-rich ash, ready to germinate with the first rainfall. A similar succession of burning of the understorey and seed falling—although on a cycle perhaps measured in hundreds of years—has produced the magnificent forests of mountain ash in southeastern Australia.

5 Sattaur, Omar 1988 Native Is Beautiful. *New Scientist*, 2 June:54–57.

FURTHER READING
Gill, A. M., Groves, R. H. and Noble, I. R. (Editors) 1981 *Fire and the Australian Biota*. Australian Academy of Science, Canberra.

Index

CHINA

Vegetables	Fruit
Soybean	Orange
Rice	Lemon
Tea	Kumquat
Millet	Persimmon
Endive	Lichi
Chinese cabbage	

Reindeer

Dog

Duck

Pig

Silkworm

Goldfish

Cormorant

Elephant

Peacock

Goose

PHILIPPINES – NEW GUINEA

Coconut
Sugarcane
Taro
Breadfruit
Banana
Sago

SOUTHEAST ASIA

Vegetables

Yam
Bamboo
Lotus
Water chestnut
Sugar palm

Fruit

Jackfruit
Durian
Rambutan
Mangosteen
Starfruit
Mango

Spices

Cloves
Nutmeg
Ginger
Turmeric

Other

Cotton
Tapa tree
(Mulberry)

Water buffalo

Chicken

Bali cattle
(Banteng)

Pacific

Ocean

Animals and plants domesticated in the Pacific Basin